GRISWOLD'S

RAILROAD ENGINEERS'

POCKET COMPANION

For the Field.

COMPRISING

RULES FOR CALCULATING DEFLEXION DISTANCES AND ANGLES, TANGENTIAL DISTANCES AND ANGLES, AND ALL NECESSARY TABLES FOR ENGINEERS;

ALSO,

THE ART OF LEVELING FROM PRELIMINARY SURVEY TO THE CONSTRUCTION OF RAILROADS, INTENDED EXPRESSLY FOR THE YOUNG ENGINEER

TOGETHER

WITH NUMEROUS VALUABLE RULES AND EXAMPLES.

BY W. GRISWOLD.

PHILADELPHIA:
HENRY CAREY BAIRD,
INDUSTRIAL PUBLISHER,
No. 406 WALNUT ST.
1866.

PREFACE.

In offering this book to the patronage of the Assistant Engineer, I would wish to remark, that the book is composed of notes that I have long been collecting.

Every Engineer has his private book of Rules, that should he want for memory in the field, he has only to refer to his book.

This book I have intended for the same object.

For the benefit of the young Engineer, I have inserted the art of leveling, running levels (as it is termed) in plain language, from preliminary surveys to the construction of a railroad; the manner of taking cross sections of the road bed, and setting slope stakes, with its rules. Every feature of the book has a tendency to attract the attention of both the Assistant and the young Engineer.

Tables that would be more used by the Assistant Engineer, are inserted, which can be relied upon as correct, as I have taken them from reliable authors.

In the art of leveling, I have made no allowance for the Earth's curvature, as in practice upon railroads no allowance is ever made.

CONTENTS.

ORDINATES.

TANGENTIAL ANGLES.

TRIGONOMETRY.

SURVEYING.

THE ART OF LEVELING.

MISCELLANIES.

TABLES.

RAILROAD ENGINEER'S

POCKET COMPANION

FOR THE FIELD.

EXPLANATION OF LETTERS AND TERMS.

P. C. Point of curve. *E. C.* End of curve.
T. P. Tangent point.
1 Station is equal to 100 feet.
A Plus Station is any number of feet less than 100.
B. S. Back sight. *F. S.* Fore sight.
Int. S. Intermediate sight. *H. Ins.* Height of instrument.

CURVATURE.

To find the radius of curves, from the deflexion angles, from chord to chord. (Chord 100 ft.)

RULE 1.

As angle of deflexion
Is to the length of the chord,
So is 57.3 to radius.

To find radius of curves, from the deflexion distance, from chord to chord. (Chord 100 ft.)

RULE 2.

The square of the chord, divided by the deflexion distance.

EXAMPLE.—Deflexion distance = 3.$\frac{49}{100}$ ft.;
Square of chord 10,000;
3.49)10,000(2865, radius.

To find the radius of a curve, from the deflexion angle, on chord of 100 ft.

RULE 3.

Divide the radius of a one degree curve (5,730) by the degrees of deflexion of 100 ft.

To find radius of a segment of a circle.

RULE 4.

Square of half the chord, added to the square of versed sine, = square of chord of half the arc; and square of chord of half the arc, divided by versed sine, = diameter, $\frac{1}{2}$ = radius.

When the angle at vertex is given, and radius, to find the tangent point.

RULE 5.

Multiply nat. tangent of half the whole angle in the curve, by radius of curve; will equal distance from vertex to tangent point.

When the distance from vertex to tangent point, and angle given, to find radius.

RULE 6.

Subtract the angle *a b c*, (Fig. 1) which is half the angle *a b d*, from 90°; the remainder will be the angle *b c a*.

Then say: As nat. sine of *b c a* is to nat. sine of *a b c*, so is *a b* or *b d* to the radius.

Having given the angle *a b d*, (Fig. 1) it is required to find the point *a* or *d*, at which to commence a curve, of a given radius.

RULE 7.

Subtract half the angle *a b d* from 90°, the remainder will be the angle *b c a* or *b c d*; then take the natural tangent of *b c a* or *b c d*, and multiply it by the given radius; the product will be *b a* or *b d*.

Having the given radius (Fig. 1) *a c*, or deflexion angle for 100 ft., of a curve, and the angle *a b d*, it is required to find the number of chords of 100 ft. that will constitute the curve.

RULE 8.

Subtract the angle *a b d* from 180°, and divide the remainder by the angle of deflexion in 100 ft.

Having the angle at vertex (Fig. 1) *e b d*, (which is the number of degrees in the curve,) and deflexion angle for 100 ft., to find the number of chords in that curve.

RULE 9.

Divide the number of degrees at vertex by deflexion angle.

With the distance *a b* or *b d*, and radius *a c* given, to find the distance from *f* to *b*, (Fig. 1.)

RULE 10.

The square of the distance from vertex to *P. C.* divided by twice the radius.

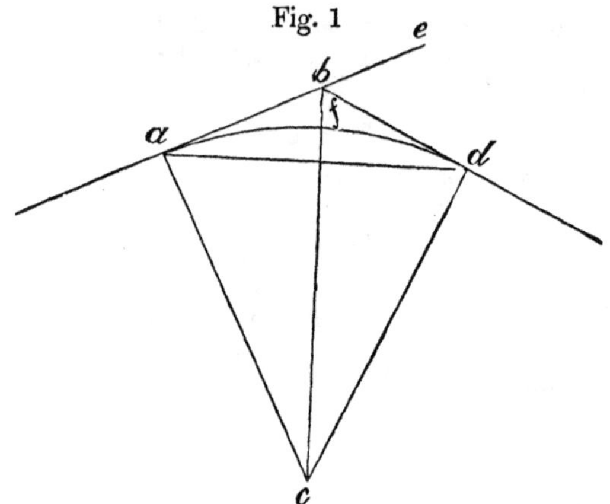

EXAMPLE.—Suppose the distance from *a* to *b* = 511 ft., and radius = 1,910 ft.,

Then $511^2 = 261{,}121 \div 2 \times 1{,}910 = 68.2$, Answer.

NOTE.—The square of any distance, divided by twice radius, will equal the distance from tangent to curve, very nearly.

When the distance *a b* or *b d* is given, and distance *f b*, (Fig. 1) to find radius.

RULE 11.

Divide the square of the distance *a b* or *b d* by the distance *f b*, equals twice the length of radius, $\frac{1}{2}$ = radius.

EXAMPLE.—Suppose the distance *f b* equals 68.2, and *a b* or *b d* equal 511.

STATEMENT.— $511^2 = 261{,}121 \div 68.2 = 3{,}820 \div 2 = 1{,}910$, radius.

To find the radius corresponding to any given angle of deflexion, and to equal chords of any given length.

RULE 12.

Subtract the angle of deflexion from 180°; then say: as nat. sine of angle of deflexion, is to nat. sine of half the remainder, so is the given chord to the radius required.

EXAMPLE.— Let the angle of deflexion be 2°, and the chord 100 ft., required the radius.
Then $2° — 180° = 178° \div 2 = 89°$.

N. S. 2°. N. Sine 89°. Chord.
STATEMENT.— 0.0349 : .999848 :: 100 : 2865, radius.

To find the circumference of a circle, when the diameter is given, or the diameter, when the circumference is given.

RULE 13.

Multiply the diameter by 3.1416, equals circumference; or, divide the circumference by 3.1416, equals diameter.

2d. As 7 is to 22,
So is the diameter to the circumference.
Or, as 22 is to 7,
So is the circumference to the diameter.

To find the length of an arc or circle, containing any number of degrees.

RULE 14.

Multiply the number of degrees in the given arc,

by 0.0087266, and the product by the diameter of the circle.

NOTE.—The circumference of a circle, whose diameter is 1, is 3.1416; it follows, that if 3.1416 be divided by 360°, the quotient will be the length of an arc of 1 degree, = 0.0087266.

REMARK.—When the arc contains degrees and minutes, reduce the minutes to a decimal of a degree.

To find the length of any arc of a circle.

RULE 15.

Subtract the chord of the whole arc from 8 times the chord of half the arc, and $\frac{1}{3}$ of the remainder is the length of the arc, nearly.

When the chord of the arc, and the chord of half the arc, are given.

RULE 16.

From the square of the chord of half the arc, subtract the square of half the chord of the entire arc; will equal the square of the versed sine; extract the square root; will equal versed sine—the versed sine and the chord given—the square of $\frac{1}{2}$ the length of the chord, added to the square of the versed sine, and square root of the remainder, will equal chord of $\frac{1}{2}$ the arc; multiply the remainder by 8, subtract the chord of the whole arc, and divide by 3, equals length of arc.

To find the circumference of an ellipses.

RULE 17.

Half the sum of the two diameters, multiplied by 3.1416; the product will equal circumference.

TABLE OF RADII — CHORDS 100 FT.

Angle of Deflexion.	Radius in Feet.	Angle of Deflexion.	Radius in Feet.	Angle of Deflexion.	Radius in Feet.	Angle of Deflexion.	Radius in Feet.
° ′		° ′		° ′		° ′	
0 15	22920	8 15	695·1	16 15	353·8	24 15	238·0
30	11460	30	674·6	30	348·4	30	235·6
45	7640	45	655·5	45	343·3	45	233·3
1	5730	9	637·3	17	338·3	25	231·0
15	4584	15	620·2	15	333·7	15	228·7
30	3820	30	603·8	30	328·7	30	226·5
45	3274	45	588·4	45	324·8	45	224·3
2	2865	10	573·7	18	319·6	26	222·3
15	2547	15	559·7	15	315·2	15	220·6
30	2292	30	546·4	30	311·0	30	218·0
45	2084·0	45	533·8	45	306·9	45	216·0
3	1910	11	521·7	19	302·9	27	214·2
15	1763	15	510·1	15	299·4	15	212·2
30	1637	30	499·1	30	295·3	30	210·3
45	1528	45	488·5	45	291·5	45	208·5
4	1433	12	478·3	20	287·9	28	206·7
15	1348	15	468·7	15	284·4	29	199·7
30	1274	30	459·3	30	280.9	30	193·2
45	1207	45	450·3	45	277·6	31	187·1
5	1146	13	441·7	21	274·4	32	181·4
15	1092	15	433·4	15	271·1	33	176·0
30	1042	30	425·5	30	268·0	34	171·0
45	996·8	45	417·7	45	265·0	35	166·3
6	955·4	14	410·3	22	262·0	36	161·8
15	917·0	15	403·1	15	260·0	37	157·6
30	882·0	30	396·2	30	257·4	38	153·6
45	849·3	45	389·6	45	254·6	39	149·8
7	819·0	15	383·1	23	250·8	40	146·2
15	790·8	15	376·9	15	248·1	45	136·5
30	764·5	30	370·8	30	245·5		
45	739·9	45	365·0	45	242·9		
8	716·8	16	359·3	24	240·5		

To find the radius corresponding to any given angle of deflexion, and to equal chords of any given length.

RULE 18.

Subtract the angle of deflexion from 180°; then say: as nat. sine of angle of deflexion is to nat. sine of half the remainder, so is the given chord to the radius required.

Fig. 2.

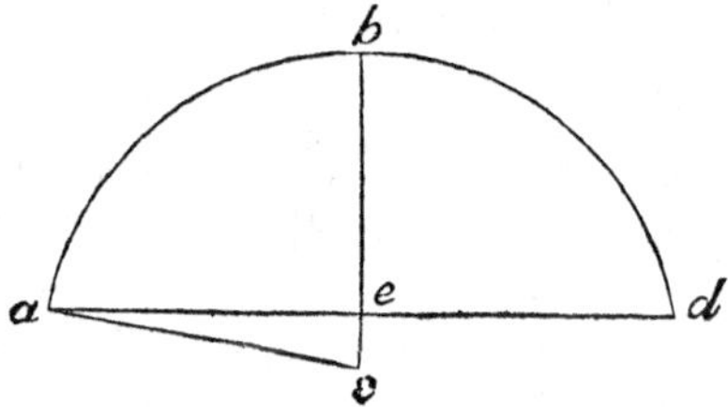

With the chord and versed sine given, to find the radius.

RULE 19.

The square of half the chord divided by versed sine; to which add the versed sine, and divide by 2.

EXAMPLE.—Suppose we have an arc (Fig. 2) with a chord *a d* of five feet, and versed sine *e b* two feet, what is the radius *a c*?

STATEMENT.—$5 \div 2 = 2{\cdot}5^2 = 6{\cdot}25 \div 2 = 3{\cdot}125 + 2 = 5{\cdot}125 \div 2 = 2{\cdot}5625$, radius *a c*.

We have two tangents with their courses given. We wish to unite those tangents, with a curve of a given radius.

Suppose we have a tangent whose course is *N.* 30° *E.*, (Fig. 3) which we wish to unite with a 3° curve to a tangent whose course is *S.* 50° *E.* We here find we have the difference of courses to be 100°. According to Rule 8, page 7, and Rule 20, page 14, we have $3,333\frac{33}{100}$ ft. to run. We start on the first given tangent, and run $3,333\frac{33}{100}$ ft. If our curve does not form a tangent of the line *g e*, but touches the point *c*, we measure in a line the same course of the first tangent, *N.* 30° *E.*, to its intersection, which distance you measure backward or forward for *P. C.*

Fig. 3.

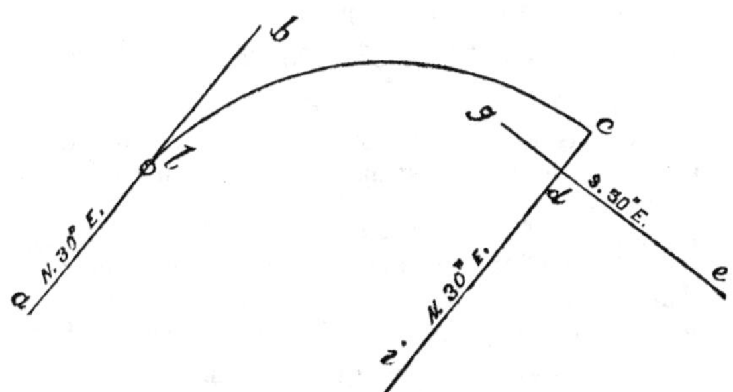

Example.—We have the tangent *a b*, *N.* 30° *E.*, and tangent *e g*, *S.* 50° *E.*; we wish to join those two tangents with a 3° curve.

Statement.—*T. N.* 30° *E.* and *T. S.* 50° *E.* = 100°, angle of deflexion, which makes 100° in the curve; consequently, the number of feet in the curve (chords 100 ft.) = 100° ÷ 3 = 33 stations and $33\frac{33}{100}$ ft. = $3{,}333\frac{33}{100}$ ft. We start from the tangent *a b*, at the point *l*, and run $3{,}333\frac{33}{100}$ ft., turning off for a 3° curve, and find, when arriving at our tangent, we are 250 ft. from the line, as *d c*; we then return and measure the same distance on the tangent *a b* from *l.* to *P. C.*

With the angle at vertex given, and degree of curvature, to find how many feet constitutes the curve.

RULE 20.

Divide the number of degrees at vertex, by degree of curvature.

EXAMPLE.—We have the angle at vertex = 100°, and degree of curvature = 3°.

STATEMENT.—100° ÷ 3° = 33 stations and $33\frac{33}{100}$ ft. = $3{,}333\frac{33}{100}$ ft.

If there is any number of feet less than 100 ft., in your curve, to find how many degrees and minutes to turn off.

RULE 21.

Say: as 100 ft. is to the number of feet you wish to run, so is the number of degrees and minutes to the number of degrees and minutes you wish to find.

EXAMPLE.—Suppose you turn off in every 100 ft., 1° 30′, how much will it be necessary to turn off in 33 ft.?

STATEMENT.—100 : 33 :: 1° 30′ : 30′, Answer.

Suppose you have a less number of degrees and minutes, than you turn off in 100 ft., to find the number of feet necessary to measure.

RULE 22.

As the whole number of degrees and minutes is to the number of degrees and minutes you wish to turn off, so is the chord, 100 ft., to the number of feet required.

EXAMPLE.—We turn off in 100 ft. 1° 30′, we wish to find the number of feet to measure for 30′.

STATEMENT.—1° 30′ : 30′ :: 100 : 33, Answer.

Fig. 4.

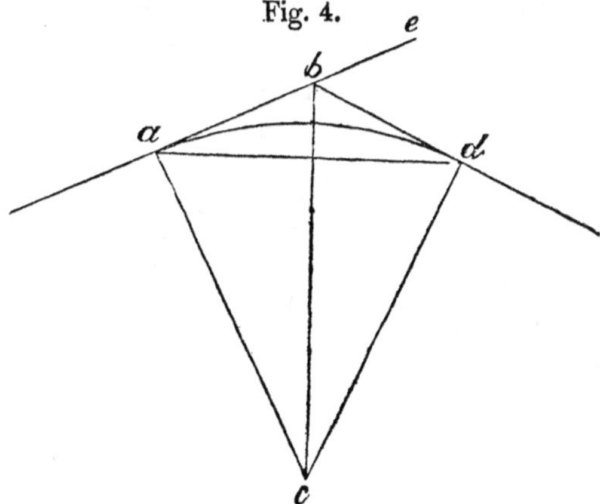

With the angle *a b d*, (Fig. 4) and distance *a b* given, to find radius *a c*.

RULE 23.

Half the angle *a b d* taken from 90° = angle *a c b*. Then say: as nat. sine of *a c b* is to nat. sine of *a b c*, so is the given side *a b* or *b d* to the radius *a c*.

NOTE.—This rule is often used in compounding curves, where the curve you have run does not fit the tangent, by measuring from a given point, on a tangent, to curve already run, to the tangent you wish to connect, as *b l* (Fig. 5) and measuring the angle *c d l*, and proceed according to rule 23.

Thus in Fig. 5, we run our curve to *n*, and find that it does not come in tangentially to the tangent *b l*, therefore to save loss of time, in long curves, we retrace our steps to the point *c*, and measure tangentially to curve *a c*, as *c d* to the tangent *b l*, and measure the angle *c d l*, and form a new radius, as *c g* or *e g*.

Fig. 5.

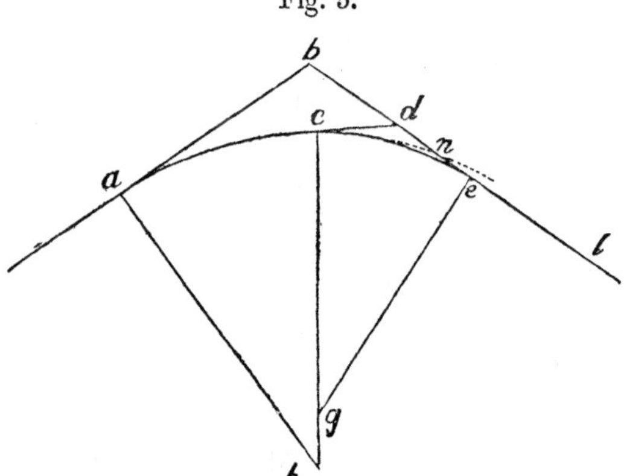

We oftentimes have two different angles in a line, and in such close proximity that it is required to put in reversed curves (Fig. 6) that will connect with the greatest possible radius; we then wish to find the greatest radius *a h* and *e g* that will connect the tangents with a reversed curve.

Fig. 6.

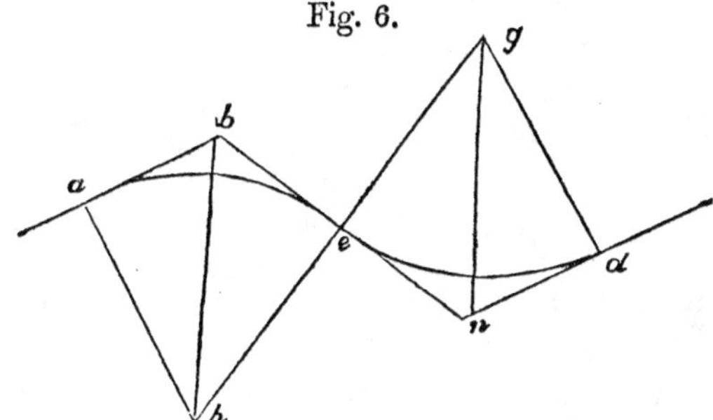

RULE 24.

Half the angle *a b e* taken from 90° leaves the angle *b h e* or *a h b*, and half the angle *b n d* taken from 90° leaves the angle *e g n*. From the table of nat. tangent take the nat. tangent of *b h e* or *a h b* and nat. tangent of the angle *e g n* and add them together.

Then say: as the sum of these two nat. tangents is to the nat. tangent of *b h e*, so is the distance *b n* to *b e*.

Again, in the triangle *b h e*, as the nat. sine of the angle *b h e*, opposite the given side *b e* just found, is to the nat. sine of the angle *h b e*, opposite the required side *h e*, so is *b e* to *h e*, the radius required.

EXAMPLE.—Let the angle *a b n* be 71° 40′, the angle *b n d* 129° 15′, and the distance *b n* be 950 ft., what is the length of the radius *h e* or *e g*?

STATEMENT 1st.—$71° 40' \div 2 = 35° 50' - 90 = 54° 10'$.

$129° 15' \div 2 = 64° 37\frac{1}{2}' - 90 = 25° 22\frac{1}{2}'$.

Nat. tangent of $54° 10' = 1·3848$.

Nat. tangent of $25° 22\frac{1}{2}' = 0·4743 + 1·3848 = 1·8591$.

Sum. N. tang. 54° 10′ *b n* *b e*
Then as 1·8591 : 1·3848 : : 950ft. : 707·63 ft.

Again, 950 — 707·63 = 242·37 ft. = *e n*.

Sine *b h e*. N. sine *h b e*.
STATEMENT 2d.—0·8107 : ·5854 : : 707·63 : 510·97, radius required.

Hence, we have the distance *h e* and *e g* = 510·97 ft., and distance *b e* 707·63 ft., and distance *e n* 342·37 ft., and degree of curvature equal to 11° 16′ for chords of 100 f.

The radius of a curve is always at right angles with its tangent.

When running curves with a compass, or transit instrument, always turn off on the vernia, one-half the number of degrees as contains degrees of curvature; because when running with an instrument, you are running tangential angles instead of deflexion angles, and the tangential angle equals one-half the deflexion angle.

ORDINATES.

To find ordinates on chords of 100 ft.

RULE 1.

The product of the segment, divided by twice the radius.

EXAMPLE.—Suppose the radius = 2·865 ft.; what would be the ordinate for 25 ft.? (Chords 100 ft.)

$25 \times 75 \div 5{,}730 = 0·327$.

Rule for getting the ordinate for 50 ft. and 25 ft. approximately.

RULE 2.

For an ordinate of 50 ft., divide the deflexion distance for 100 ft. by 8. For 25 ft. three-fourths of the ordinate for 50 ft.

EXAMPLE.—Suppose the deflexion = 3°, which deflexion distance = 5·235 ÷ 8 = 0·654, ordinate for 50 ft., and three-fourths of 0·654 = 0·491, ordinate for 25 ft.

To find the middle ordinate to any given radius, and to any given chord.

RULE 3.

From the square of the radius, subtract the square of half the chord, and take the square root of the remainder from the radius = middle ordinate.

EXAMPLE.—What is the length of the middle ordinate *d e*, (Fig. 1) the radius *a c* being 2·5625, and chord *a b* 5 ft.

STATEMENT.—$2·5625^2 - 2·5^2 = \sqrt{·316406} = ·5625 - 2·5625 = 2$ ft., middle ordinate.

TABLE OF ORDINATES — CHORDS 100 FT.

Angle of Deflexion.		Length of Ordinates in Feet.									
°	′	50	45	40	35	30	25	20	15	10	5
0	5	·018	·018	·017	·016	·015	·014	·012	·009	·006	·003
	10	·036	·036	·035	·033	·031	·027	·023	·019	·013	·007
	15	·054	·054	·052	·049	·046	·041	·035	·028	·019	·010
	20	·073	·072	·070	·066	.061	·055	·047	·037	·026	·014
	25	·091	·090	·087	·082	·076	·068	·058	·046	·032	·017
	30	·109	·108	·105	·099	·092	·082	·070	·055	·039	·020
	35	·127	·126	·123	·116	·108	·096	·082	·065	·045	024
	40	·145	·144	·140	·133	·123	·110	·093	·074	·052	·027
	45	·163	161	·157	·149	·137	·123	·105	·083	·058	·031
	50	·182	·180	·175	·166	·153	·138	·117	·092	·065	·034
	55	·200	·198	·192	·182	·168	·151	·128	·102	·071	·038
1		·218	·216	·209	·198	·183	·164	·140	·111	·078	·041
	5	·236	·234	·226	·215	·198	·178	·152	·120	·085	·044
	10	·254	·252	·244	·231	·214	·191	·163	·130	·091	·048
	15	·273	·270	·261	·248	·229	·205	·175	·139	·098	·051
	20	·291	·288	·279	·264	·244	·218	·187	·148	·104	·055
	25	·309	·306	·296	·281	·259	·232	·198	·157	·111	·058
	30	·327	·324	·314	·297	·275	·246	·210	·167	·117	·062
	35	·345	·342	·331	·314	·290	·259	·221	·176	·124	·065
	40	·364	·360	·349	·330	·305	·273	·233	·185	·130	·069
	45	·382	·378	·366	·347	·321	·287	·245	·195	·137	·072
	50	·400	·396	·384	·364	·336	·300	·256	·204	·144	·076
	55	·418	·414	·401	·380	·351	·314	·268	·213	·150	·079
2		·436	·432	·419	·397	·366	·327	·280	·222	·157	·083
	5	·454	·450	·436	·413	·382	·341	·291	·232	·163	·086
	10	·473	·468	·454	·430	·397	·355	·303	·241	·170	·089
	15	·491	·486	·471	·446	·412	·368	·315	·250	·176	·093
	20	·509	·504	·489	·463	·428	·382	·326	·260	·183	·096
	25	·527	·522	·506	·480	·443	·396	·338	·269	·190	·100
	30	·545	·540	·524	·496	·458	·409	·350	·278	·196	·103
	35	·564	·558	·541	·513	·474	·423	·361	·288	·203	·107
	40	·582	·576	·559	·529	·489	·436	·373	·297	·209	·110
	45	·600	·594	·576	·546	·504	·450	·384	·306	·216	·114
	50	·618	·612	·594	·562	·519	·464	·396	·315	·222	·117
	55	·636	·630	·611	·579	·535	·477	·408	·325	·229	121
3		·654	·648	·629	·595	·550	·491	·419	·334	·235	·124
	5	·673	·666	·646	·612	·565	·504	·431	·343	·242	·128
	10	·691	·684	·664	·629	·581	·518	·443	·353	·249	·131
	15	·709	·702	·681	·645	·596	·532	·454	·362	·255	·134
	20	·727	·720	·699	·662	·611	·545	466	·371	262	·138
	25	·745	·738	·716	·678	·627	·559	·478	·380	·268	·141
	30	·764	·756	·734	·695	·642	·573	489	·390	·275	·145
	35	·782	·774	·751	·711	·657	·586	·501	·399	·281	·148
	40	·800	·792	·769	·728	·673	·600	·512	·408	·288	·152
	45	·818	·810	·786	·744	·688	·613	·524	·418	·294	·155
	50	·836	·828	·804	·761	·703	·627	·536	·427	·301	·159
	55	·854	·846	·821	·778	·718	·641	·547	·436	·308	·162
4		·873	·864	·839	·794	·734	·654	·559	·445	·314	·166

TABLE OF ORDINATES — CHORDS 100 FT.

Angle of Deflexion.	Length of Ordinates in Feet.									
	50	45	40	35	30	25	20	15	10	5
4° 15′	·927	·918	·891	·844	·780	·695	·594	473	334	·176
30	·981	·972	·944	·893	·825	·736	·629	·501	·354	·186
45	1·036	1·026	·996	·943	·871	·777	·664	·529	·373	·196
5	1·091	1·080	1·048	·993	·917	·818	·699	·557	·393	·207
15	1·146	1·134	1·100	1·042	·963	·859	·734	·585	·413	·217
30	1·200	1·188	1·153	1·092	1·009	·900	·769	·613	·432	·228
45	1·255	1·242	1·205	1·141	1·055	·941	·804	·640	·452	·238
6	1·309	1·296	1·258	1·191	1·100	·982	·839	·668	·472	·249
30	1·419	1·404	1·362	1·290	1·192	1·064	·909	·724	·511	·269
45	1·473	1·458	1·415	1·339	1·238	1·105	·944	·752	·531	·280
7	1·528	1·512	1·467	1·389	1·284	1·146	·979	·779	·551	·290
30	1·637	1·620	1·572	1·488	1·375	1·228	1·048	·835	·590	·311
8	1·746	1·728	1·677	1·587	1·467	1 310	1·118	·891	·629	·332
30	1·855	1·836	1·782	1·687	1·559	1·392	1·188	·946	·669	·353
9	1·965	1·944	1·886	1·787	1·651	1·474	1·258	1·002	·708	·373
30	2·074	2·052	1·991	1·887	1·742	1 556	1·328	1·057	·748	·394
10	2·183	2·161	2·096	1·987	1·834	1·637	1·398	1·114	·787	·415
30	2·292	2·269	2·201	2·087	1·926	1·719	1·468	1·170	·827	·436
11	2·401	2 377	2·306	2·186	2·018	1·802	1·538	1·226	·866	·457
30	2·511	2·486	2·411	2·236	2·110	1·884	1·609	1·282	906	·478
12	2·620	2·594	2·516	2·386	2·203	1·967	1·680	1 339	·946	·499
13	2·839	2·811	2·726	2·585	2·387	2·132	1·820	1·451	1·025	·541
14	3·058	3·028	2·937	2·785	2·571	2·297	1·961	1·564	1·105	·583
15	3·277	3·245	3·147	2·984	2·756	2·462	2·102	1·676	1·184	·625
16	3·496	3·462	3·358	3 184	2·941	2·627	2·243	1·789	1·264	·667
17	3·716	3·680	3·569	3·384	3·125	2·792	2·384	1·902	1·344	·709
18	3·935	3·897	3·779	3·584	3·310	2·958	2·525	2.014	1·424	·751
19	4·155	4·115	3·990	3·784	3·495	3·123	2·666	2·127	1·504	·793
20	4·375	4·332	4·201	3·984	3 680	3·288	2·808	2·240	1·583	·836
21	4·595	4·549	4·412	4·184	3·864	3·454	2·950	2·353	1·663	·879
22	4·815	4·768	4·624	4·386	4 050	3·620	3·093	2·467	1·744	·922
23	5·035	4·986	4 836	4·587	4·237	3·786	3 236	2·581	1·824	·965
24	5·255	5·204	5·048	4·789	4·423	3·952	3·379	2·695	1·905	1·008
25	5 476	5·422	5·260	4·989	4·609	4 119	3·522	2·809	1 986	1 051
26	5 697	5·642	5·473	5·192	4·798	4·286	3·665	2·924	2·068	1·094
27	5 918	5·860	5·685	5·393	4·984	4·454	3·808	3·039	2·150	1·137
28	6·139	6·079	5·898	5·595	5·171	4·622	3·952	3·154	2·232	1·181
29	6·361	6 298	6·110	5·796	5·357	4·790	4·095	3·269	2·314	1·224
30	6·582	6·517	6·323	5·999	5·544	4·958	4·239	3·385	2·396	1·268
31	6·804	6·737	6 537	6·202	5·733	5 127	4·384	3·502	2·481	1·312
32	7·027	6·957	6 751	6·406	5·922	5·297	4 530	3·619	2·565	1 356
33	7·249	7·178	6·965	6·609	6·111	5·467	4·676	3·737	2·649	1·401
34	7·472	7·398	7·179	6·813	6 300	5·637	4·822	3·854	2·733	1·445
35	7·694	7·619	7·393	7·017	6·489	5·807	4·968	3·972	2·817	1·490
36	7·918	7·841	7·609	7·222	6·679	5·978	5 115	4 090	2·901	1·535
37	8·143	8 063	7·825	7·427	6·870	6·149	5·262	4·209	2·985	1·581
38	8·367	8·286	8·041	7·633	7·060	6·320	5·410	4·327	3·069	1·626

RULE 4.

Subtract the tabular cosine of the tangential angles from 1, and multiply the remainder by the radius. (Chords 100 ft.)

EXAMPLE.—Radius 819 ft., angle of deflexion would be 7° to chord of 100 ft; what will be the length of the middle ordinate?

STATEMENT.—Here tabular cosine of the tangential angle $3\frac{1}{2}° = \cdot 998135$, which, subtracted from $1 = \cdot 001865$, which, multiplied by radius, 819 ft., = ordinate, 1·528.

Fig. 1.

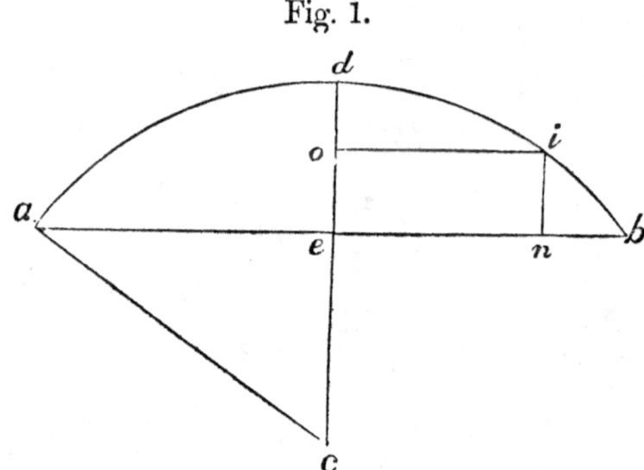

Having the middle ordinate *d e*, (Fig. 1) it is required to find any other ordinate, as *i n*.

RULE 5.

Subtract the middle ordinate *d e*, from the radius *a c*, the remainder will be *e c*; from the square of the radius *a c*, subtract the square of the distance *o i* or *e n*, and extract the square root of the remainder; this square root will be *o c*; subtract *e c* from *o c*; the remainder will be *o e*, which is equal to *i n*, the required ordinate.

Example.—The middle ordinate *d e* (Fig. 1) of a 100 ft. chord *a b*, to a radius of 819 ft. = 1·52; it is required to find the length of the ordinate *i n* 20 ft. from the middle one *d e*.

Statement.—819 — 1·528 = 817·472.

Again, $819^2 = 670761 - 20^2 = \sqrt{670361} = 818{\cdot}756 = o\ c - 817{\cdot}472 = 1{\cdot}284$, will equal *o e* or *i n*, the required ordinate.

To find middle ordinate approximately. Chord 100 ft.

RULE 6.

Multiply the ordinate of a 1° curve by the deflexion angle of 100 ft. This rule is sufficiently close for curves of not less than 500 ft. radius.

2d. Multiply the chords together, and divide by twice the radius.

Fig. 2.

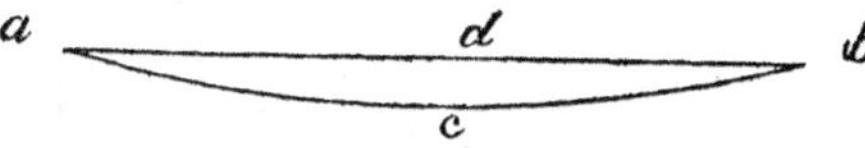

To find the ordinate for a railroad bar *a c b* (Fig. 2) 24 ft. long.

RULE 7.

Multiply one-half the length of the rail by one-fourth its length, and divide by the radius.

EXAMPLE.—Rail 24 ft. long, radius 5730.

STATEMENT.—$24 \div \frac{1}{2} = 12 = \frac{1}{2}$ the length of rail.
$24 \div \frac{1}{4} = 6 = \frac{1}{4}$ the length of rail.
Then $6 \times 12 = 72 \div 5730 = 0{\cdot}01$, ordinate required.

2d. Take one-fourth of the square of the length of the rail, and divide it by twice the radius.

An approximate rule for calculating the middle ordinate of a sub-chord, when the middle ordinate is given. (Chord 100 ft.)

RULE 8.

As the square of the length of the whole chord is to the square of the length of the sub-chord, so is the middle ordinate of the chord to the middle ordinate of the sub-chord.

EXAMPLE.—Chord 100 ft., middle ordinate ·218, what will be the middle ordinate of the sub-chord 50 ft.?

STATEMENT.—$100^2 : 50^2 :: {\cdot}218 : {\cdot}0545$.

Fig. 3.

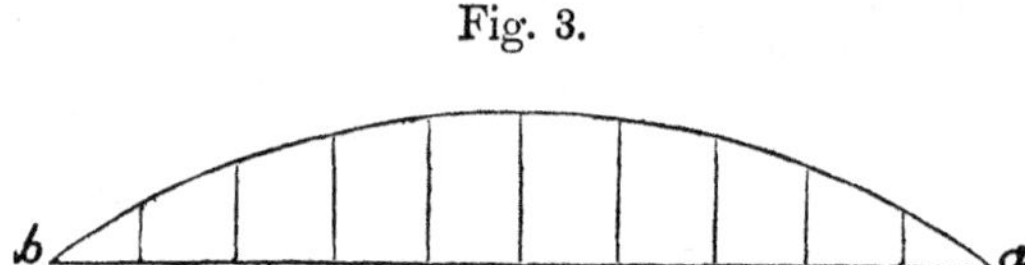

In running a curve for track, after the grading is done, it is necessary to put in intermediate ordinates, if the curve exceeds 1°; (Fig. 3) these intermediates are from 10 to 20 ft. apart, and instead of running these intermediates with an instrument, the best method is, after your points are put in with an instrument 100 ft. apart, to draw a small cord or twine, as *a b*, and measure off your ordinates with a graduated rod, or with the leveling rod.

NOTE.— Refer to Rule 1, page 19 for intermediates or to table.

When the chord and radius are given, to find middle ordinate. (Chord 100 ft.)

RULE 9.

Divide the square of the chord by eight times the radius.

DEFLEXION DISTANCE.

To find the deflexion distance for 100 ft., with any given radius.

RULE 1.

The square of the chord divided by the radius.

RULE 2.

Divide the constant number 10,000 (chords 100 ft.) by the radius in feet, equals deflexion distance.

To find the deflexion distance for any given radius for chords of 100 ft.

RULE 3.

Divide the given chord by radius, will give the nat. sine of the deflexion angle, which, multiplied by the chord, will equal the required distance.

NOTE.—The tangential distance for 100 ft. is equal to one-half the deflexion distance, and the tangential angle is always equal to one-half the deflexion angle.

Fig. 1.

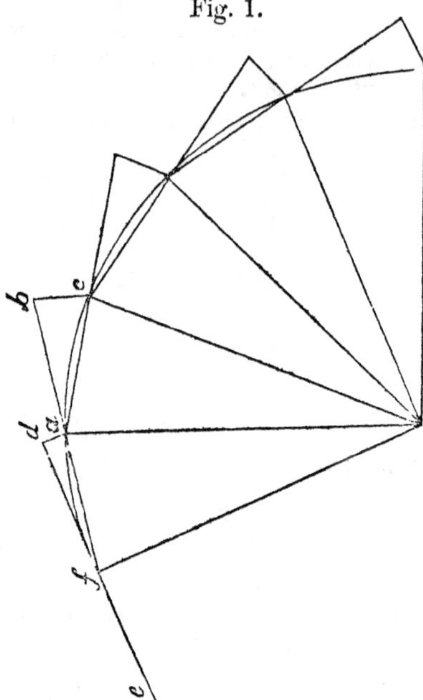

In putting in curves by deflexion distances it is quite necessary, for accuracy, to measure both on the line of deflexion and chord of the arc, as *a b* and *a c*, (Fig. 1) by first finding your point *b*, and swinging your chain to *c*, and measuring your deflexion distance on the line *b c*.

In commencing your curve on the tangent *e d* you

measure your distance *f d*, and lay off *a d* equal to one-half *b c*, or one-half the deflexion distance, as *a d* is the tangential distance, and the tangential distance is equal to one-half the deflexion distance.

If you wish to put in intermediates, as it frequently occurs, at the end of a curve.

RULE 4.

Find your deflexion distance for 100 ft., string a line, and put in the required ordinate.

To find the deflexion distance for any number of feet less than 100.

RULE 5.

Take the deflexion distance for 100 ft. and multiply it by the required chord, and divide the product by the length of the whole chord, 100 ft., and subtract the ordinate corresponding in feet and degree.

EXAMPLE.—We wish to get the deflexion distance for 25 ft., for a 15° curve.

STATEMENT.—Deflexion distance equals 26·11 ft., therefore:

$26{\cdot}11 \times 25 = 652{\cdot}75 \div 100 = 6{\cdot}5275.$

Now the ordinate of a 15° curve for 25 ft. = 2·462, and 6·5275 — 2·462 = 4·0655, deflexion distance for 25 ft.

NOTE.—The above Rule is sufficiently close for all practical work.

Where it is required to find the deflexion or tangential distance for more than 50 ft., subtract the distance to be found from 100, and find the ordinate corresponding to the remainder in feet.

To find the deflexion point for any number of feet, at commencement of curves and ending, where the distance is less than 100 ft.

RULE 6.

In commencing a curve, multiply the tangential distance in feet by the number of feet in the chord you wish to find, and divide the product by the length of the whole chord, 100 ft., and measure the distance on the end of a 100 ft. chord; then a line drawn from this point to the *P. C.*, the length of the chord measured you desire to find, on this line, is the point of deflexion.

RULE 7.

To end your curve: Take half the deflexion distance for 100 ft., then multiply the remaining distance (which is the tangential distance,) by the number of feet you wish to find, and divide the product by the length of the whole chord, 100 ft., and measure on the tangential distance, the distance just found; then string a line from this point to the last station given, and measure, on this last given line, the distance required, will give the *T. P.* or *E. C.*

To form a tangent to the curve: Measure as many feet more on the tangential distance, and a line drawn from the last given point to *T. P.* or *E. C.* will be the course of the tangent.

Example to Rule 6.—Suppose in commencing a curve we wish to find the deflexion distance of 25 ft. as at *r*, (Fig. 2) for a 15° curve. We take the tangential distance *c f* and multiply it by 25 ft., and divide by 100 ft., (the length of the whole chord,) will equal the distance *c s*; now if we measure in the line *b s* 25 ft. from *b* to *r*, at *r* will be the required point. We then measure 100 ft. from *r* to *d*, in line with *b r*, and measure the deflexion distance *d e*, equal to twice *c f*, for our next full station. We then measure 100 ft. on the line *r e* from *e* to the point *h*, and measure the distance *h i* for our next station, so on to the last station.

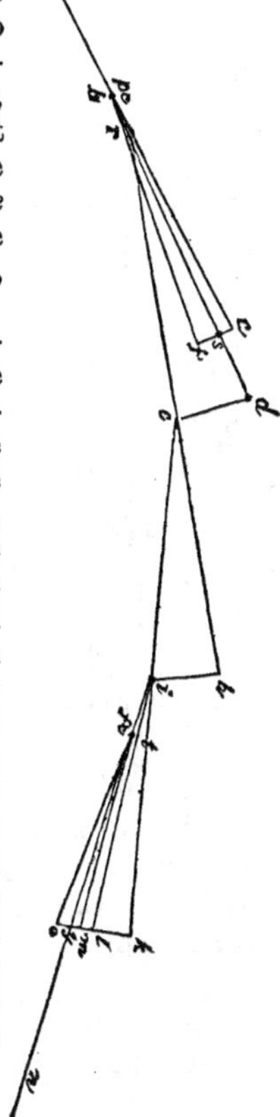

Fig. 2.

Example to Rule 7.—Suppose in ending our curve we wish to find the deflexion distance for 25 ft., to conclude the curve; (Fig. 2) suppose *i* to be the last station in the curve, and the point *t* 25 ft., to be the *E. C.*; we produce the line *e i* to *k* 100 ft. from *i*, and measure one-half the deflexion distance, *k o* = *l*, then multiply the distance *k l* by 25 ft., and divide by the length of the whole chord *i k*, 100 ft., will equal the distance *l m*, and on the line drawn from *m* to *i*, 25 ft. measured from *i* to *t*, on this line *i m*, will be the required point. To form a tangent to the point *t*, measure *m y* equal to *l m*, will form the tangent *t n* to the curve.

TABLE OF DEFLEXION DISTANCES.

Deflexion Angle.	Length of Chords in Feet.			
	100	75	50	25
° ′ 0 30	·872	0·572	0·247	0·136
45	1·308	0·899	0·491	0·204
1	1·745	1·144	0·654	0·272
30	2·618	1·717	0·982	0·408
45	3·054	2·003	1·145	0·476
2	3·490	2·290	1·309	0·545
30	4·363	2·863	1·636	0·681
45	4·799	3·419	1·799	0·749
3	5·235	3·435	1.963	0·818
30	6·108	4·008	2·290	0·954
45	6·544	4·295	2·454	1·023
4	6·980	4·581	2·617	1·091
30	7·853	5·153	2.945	1·227
45	8·289	5·440	3·108	1.295
5	8.722	5·723	3·270	1·362
6	10·470	6·870	3·926	1·635
7	12·210	8·011	4·577	1·906
8	13·950	9·152	5·229	2·177
9	15.680	10·286	5·875	2·446
10	17·430	11·435	6·532	2·720
11	19·170	12·575	7·184	2·990
12	20·940	13·738	7·850	3·268
13	22·640	14·848	8·481	3·528
14	24·370	15·980	9·127	3·795
15	26·110	17·120	9·778	4·065

If two lines vary any number of degrees, to find the distance approximately at their extremities.

RULE 8.

Say; If they vary 1·745 in 1° for 100 ft., it will vary thirty times as much in an angle of 30° for

100 ft.; if it is more than 100 ft., make a second statement.

Example.—Suppose we have an angle of 30° and 400 ft. long, what is the distance apart at their extremities?

Statement.—As 1° : 30° : : 1·745 : 52·35; 52·35 is the difference for 100 ft.

Then as 100 ft. : 400 ft. : : 52·35 : 209·4, the difference for 400 ft.

Suppose we run a line, on a given course, with the intention of striking a certain point, and find that we deviate from that point, to find the course of the second line that will unite these two points on a straight line.

RULE 9.

Multiply the difference of variation in feet by 57·3,* and divide the product by the length of the line, the quotient either added or subtracted, as necessity requires, will be the course of the line that will unite the two points together.

Example.—Suppose the difference of variation = 209·4 ft., and length of line 400 ft., and course *N.* 29° 59¾′ *E.*, what would be the course of the second line, if the point desired is *N. W.* of the line run?

Statement.— 209·4 × 57·3 ÷ 400 = 29·9965 = 29° 59¾′.

Then course *N.* 29° 59¾′ *E* — 29° 59¾′ = course *N.* 0° *E.*

* 57·3 is the radius of a circle (nearly) in such parts as the circumference contains 360.

DEFLEXION ANGLE.

To find the deflexion angle corresponding to any given radius. (Chords 100 ft.)

RULE 1.

Divide the chord by the radius; the quotient will be the natural sine of the deflexion angle; therefore, the number of degrees corresponding to this sine, in the table of nat. sines, equals the deflexion angle.

RULE 2.

The deflexion angle may be found by dividing the radius of a 1° curve, 5730, by the radius in feet, (approximately.)

To find the deflexion angle for any plus distance, or less than 100 ft.

RULE 3.

Multiply one-half the deflexion angle by the plus distance, and divide the product by 100 ft., (length of whole chord,) and add it to one-half the deflexion angle.

EXAMPLE.—Suppose we are running a 15° curve by deflexion distances; we wish to find the deflexion angle for 25 ft.

STATEMENT.—$15^\circ \div 2 = 7^\circ\ 30'$, one-half the deflexion angle.

Then, $7^\circ\ 30' \times 25 = 187^\circ\ 30' \div 100 = 1^\circ\ 52\frac{1}{2}'$.

Then, $1^\circ\ 52\frac{1}{2}' + 7^\circ\ 30' = 9^\circ\ 22\frac{1}{2}'$, deflexion angle for 25 ft.

For deflexion angles corresponding to any given radius, refer to table of radii, page 11.

TANGENTIAL DISTANCE.

To find the tangential distance for any radius, on chords of 100 ft.

RULE 1.

Divide the square of half the chord (50 ft.) by the radius, and multiply the quotient by two.

RULE 2.

Divide the square of the whole chord by twice the radius.

To find the tangential distance for any number of feet less than 100.

RULE 3.

Multiply the tangential distance for 100 ft. by the number of feet required, less than 100 ft., and divide the product by 100 ft., and from the quotient take the ordinate corresponding to the degree of curvature and feet; will equal the tangential distance for the required number of feet, less than 100 ft.

To find the tangential distance for any number of feet.

RULE 4.

Divide the square of the chord given by twice the radius.

In running curves, with equal chords on more than 100 ft., the tangential distances increase as the squares of the number of chords: thus, for 2, 3, 4, 5, 6 chords, 4, 9, 16, 25, 36, multiplied into the tangential distance of 1 chord, will equal each tangential distance respectively.

Or: the square of the length of the chord divided by twice the radius, will equal the tangential distance for any number of feet.

TABLE OF TANGENTIAL DISTANCES.

Deflexion Angle.	Length of Chords in Feet.			
	100	75	50	25
° ′				
0 30	0·436	0·245	0·109	0·027
45	0·654	0·367	0·164	0·040
1	0·873	0·490	0·218	0·054
30	1·309	0·736	0·327	0·081
45	1·527	0·858	0·381	0·095
2	1·745	0·981	0·436	0·109
30	2·182	1·227	0·546	0·136
45	2·399	1·349	0·599	0·150
3	2·618	1·472	0·655	0·163
30	3·054	1·717	0·763	0·190
45	3·272	1·841	0·818	0·205
4	3·490	1·963	0·872	0·218
30	3·927	2·209	0·982	0·246
45	4·145	2·331	1·036	0·259
5	4·361	1·452	1·089	0·272
30	4·798	2·698	1·199	0·299
45	5·015	2·820	1·252	0·312
6	5·235	2·944	1·308	0·326
7	6·105	3·432	1·524	0·380
8	6·975	3·924	1·741	0·433
9	7·840	4·406	1·955	0·486
10	8·715	4·899	2·174	0·541
11	9·585	5·387	2·391	0·594
12	10·470	5·885	2·615	0·650
13	11·340	6·373	2·831	0·703
14	12·210	6·860	3·047	0·755
15	13·080	7·348	3·263	0·808

TANGENTIAL ANGLES.

To find the tangential angle for a chord of 100 ft., with any given radius.

RULE 1.

Divide half the chord by the radius; the quotient will be the natural sine of the tangential angle; and the angle corresponding to this sine, in the table of nat. sines, is the angle required.

To find the tangential angle for any number of feet less than 100 ft.

RULE 2.

Multiply the tangential angle by the number of feet given, and divide the product by the length of the whole chord, (100 ft.)

EXAMPLE.—Suppose we have the tangential angle $= 7° 30'$, and wish to find the angle for 25 ft.

STATEMENT.—$7° 30' \times 25$ ft. $\div$ 100 ft. $= 1° 52\frac{1}{2}'$, tangential angle for 25 ft.

Sometimes in running curves it is not necessary to set points in every chord, or 100 ft., and is more expedient, as running curves on a preliminary survey. They can be put in every 2, 3, or 400 ft., as you choose. We wish to find the tangential angle for any number of chords.

RULE 3.

Multiply the tangential angle for 100 ft. by the number of chords you wish to subtend, will equal the tangential angle required.

REMARK.—In running curves, the correct way of measuring with a chain for each station, is to measure around the curve. Instead of this the chain is stretched across, forming a chord; the difference of distance is so comparatively small to a radii of 500 ft., that it is not necessary we should measure around, or make an allowance on the chain; but in running curves with long chords of 3, 4, or 500 ft., it is necessary, for accuracy, to make sufficient allowance, for which I will put in a table of long chords the lengths necessary to subtend from 1 to 4 stations.

TABLE OF LONG CHORDS.

Deflexion Angle.	Length of Chords in Feet required to Subtend.			
	1 Station.	2 Stations.	3 Stations.	4 Stations.
°				
1	100	200	300	400
2	100	200	299·9	399·7
3	100	200	299·7	399·3
4	100	199·9	299·6	398·9
5	100	199·9	299·2	398·0
6	100	199·7	298·8	397·3
7	100	199·6	298·4	396·2
8	100	199·6	298·0	395·1
9	100	199·4	297·5	394·1
10	100	199·2	297·0	392·4

To find the length of long chords.

RULE 4.

Multiply the natural sine of the tangential angle of the given chord by twice the radius.

EXAMPLE.—The tangential angle for one station = 5°, and radius = 573·7 ft; what would be the length of the chord of four stations?

STATEMENT.—The tangential angle for four stations would equal $4 \times 5° = 20°$, and nat. sine of $20° = \cdot 3420201$; twice the radius $= 1147\cdot 4 \times \cdot 3420201 = 392\cdot 4$, length of chord necessary to subtend an arc of four stations.

TRIGONOMETRY.

The angle *a* given, (Fig. 1) and hypothenuse given, to find the leg *c b*.

RULE 1.

By natural sines.—As unity, or one, is to the length of the hypothenuse, so is the natural sine of the smallest angle to the length of the shortest leg.

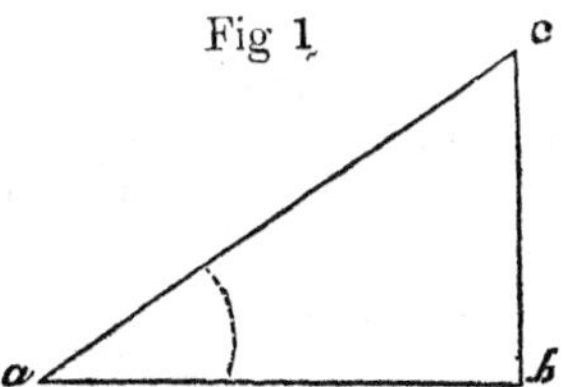

EXAMPLE.—Given the angle *b a c* 35° 30′, and hypothenuse 25 rods; to find *c b*.

STATEMENT.—1 : 25 : : 0·580703 : 14·5175.

To find the length of the leg *a b*.

RULE 2.

The difference of the sums of the squares of the legs *a c* and *c b*, and extract the square root; will equal the leg *a b*.

Example.—Given the leg *a c* 25 rods, and leg *c b* 14·5175; to find the leg *a b*.

Statement.—$\sqrt{25^2 - 14{\cdot}5175^2} = 20{\cdot}35$.

To find the leg *a b*, (Fig. 1.)

RULE 3.

By nat. sines.—As unity, or one, is to the nat. sine of the angle *a c b*, so is the hypothenuse to the leg *a b*.

Example.—Given the hypothenuse 25 rods, and angle *a c b* 54° 30′; to find the leg *a b*.

Statement.—1 : 0·8141155 : : 25 : 20·35.

The angles and leg *a b* given, (Fig. 1) to find the hypothenuse *a c*, and leg *b c*.

RULE 4.

By nat. sines.—As the nat. sine of the angle opposite the given leg *a b* is to the length of given leg, so is unity, or one, to the length of the hypothenuse.

Example.—Given the angles *a c b* 54° 30′, and *b a c* 35° 30′, and leg *a b* $20\frac{35}{100}$ rods; to find the leg *a c*.

Statement.—0·8141155 : 20·35 : : 1 : 25.

Refer to Rule 2, page 38.

To find leg *c b* by nat. sines.

RULE 5.

As the nat. sine of the angle *a c b*, opposite the given leg, is to the given leg, so is the nat. sine of the angle *b a c*, opposite the required leg, to the leg *c b*.

EXAMPLE.—Given the angle *a c b* (Fig. 1) 54° 30′, and angle *b a c* 35° 30′, and leg *a b* $20\frac{35}{100}$ rods; to find the leg *b c*.

STATEMENT.—0·8141155 : 20·35 : : 0·580703 : 14·52, leg *b c*.

The hypothenuse and one leg given; to find the angles and the other leg.

RULE 6.

By nat. sines.—The angle opposite the given leg may be found by the following proportion: As the hypothenuse is to unity, or one, so is the given leg to the nat. sine of its opposite angle.

EXAMPLE.—Given the hypothenuse *a c* 25 rods, and leg *a b* 20·35 rods; to find the angles.

STATEMENT.—25 : 1 : : 20·35 : 8141155, nat. sine of the angle *a c b*; the nearest corresponding number of degrees and minutes in the table of nat. sines gives the angle *a c b* 54° 30′, and the angle *a b c* being 90°, the angle *b a c* would be 35° 30′, because in a right-angle triangle there is always 180°.

The leg *a c* given, (Fig. 1) and angle *b a c*, to find the other leg, *a b*, by cosine.

RULE 7.

Multiply the cosine of the angle *b a c* by the hypothenuse *a c*.

EXAMPLE.—Given the hypothenuse *a c* 25 rods, and angle *b a c* 35° 30′; to find the leg *a b*.

STATEMENT.—0·8141155 × 25 = 20·35, length of the leg *a b*.

The leg *a b* found, (Fig. 1) to find the leg *b c* by nat. tangent.

RULE 8.

Multiply the base by the nat. tangent of the angle opposite the required leg.

EXAMPLE—Given the leg *a b* 20·35, and angle *b a c* 35° 30′; to find the leg *b c*.

STATEMENT.—0·713293 × 20·35 = 14·5, the required leg *b c*.

The angle *a c b*, and leg *b c*, given, (Fig. 1) to find the leg *a b*, by nat. tangents.

RULE 9.

Multiply the nat. tangent of the angle *a c b* by the leg *b c*.

EXAMPLE.—Given the leg *b c* 14·5, and angle *a c b* 54° 30′; to find the leg *a b*.

STATEMENT.—1·401948 × 14·5 = 20·35, length of leg required, *a b*.

Solution of a Right-angled Triangle.

The sine of the angle *c* equals the cosine of the angle *a*, and the sine of the angle *a* equals the cosine of the angle *c*.

The tangent of the angle *a* equals the cotangent

of the angle *c*, and the tangent of the angle *c* equals the cotangent of the angle *a*.

The leg *a b* divided by the leg *a c* equals the nat. sine of the angle *a c b*, or the nat. cosine of the angle *b a c*; the leg *b c* divided by the leg *a b* equals the nat. tangent of the angle *b a c*, or the nat. cotangent of the angle *a c b*; the leg *b c* divided by the leg *a c* equals the nat. sine of the angle *b a c*, or the nat. cosine of the angle *a c b*.

SURVEYING.

In running lines, obstructions, viz: rivers, ponds, &c., occur, by which other means have to be resorted to, besides measuring with a chain.

Fig. 2.

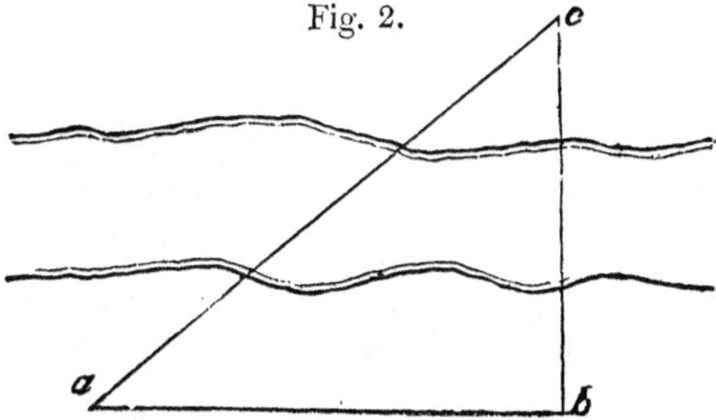

The point *c* occurs on our line *b c*, and we wish to know the distance *b c*.

RULE 10.

From *b* at right angles to the line *b c* measure any convenient distance, as *a*, and secure the point *a*; measure the angle *b a c*; then multiply the nat. tangent of the angle *b a c* by the distance *a b*; will equal the distance *b c*.

Fig. 3.

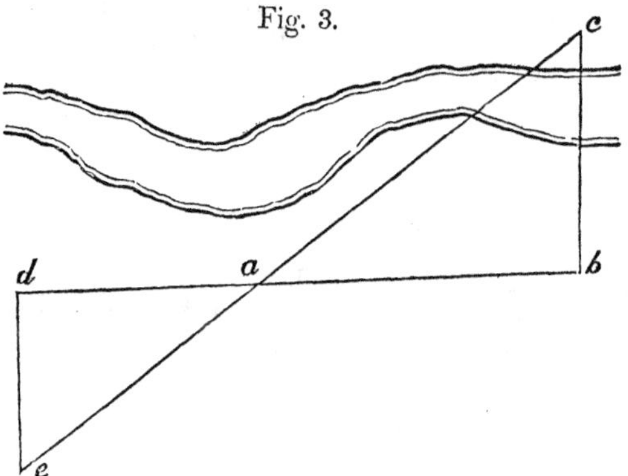

In case you should not have a book of tables of nat. tangents, the above method could be resorted to, with nearly as much accuracy as the method given in Fig. 2.

The point *c* occurring in the line, (Fig. 3) we wish to know the distance *b c*.

RULE 11.

At right angles from *b c* measure on the line *b d* to *a*, and secure the point *a*, any convenient point, and measure any convenient distance, as *d*, and at right angles describe the line *d e*, with your instrument at *a*, on the line *a c*, produce it to *e*, intersecting the line *d e*.

Then with the distance *b a*, *a d*, and *d e*, given, the distance *b c* can be found proportionately to the triangle *a d e*.

Say: As the distance *a d* is to *a b*, so is *d e* to *b c*.

EXAMPLE.— Given the distance *b a* 20 ft., distance *a d* 15 ft., and distance *d e* $11\frac{1}{2}$ ft.; to find the distance *b c*.

STATEMENT.—As $15:20::11\frac{1}{2}:15\frac{33}{100}$ ft.

Fig. 4.

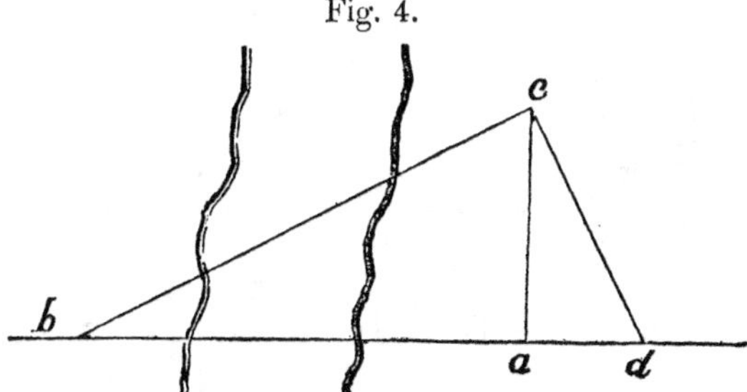

The above (Fig. 4) could be resorted to in preference to Fig. 3.

Given *b*, the inaccessible object, and *d b* part of the line of survey; we wish to find the distance from *a* to *b*.

RULE 12.

Measure on the line *a c* (at right angles with *a b*,) any convenient distance, as at *c*; then at right angles to b c run the line c d to its intersection with the line d b at d, measure the distance a d, and the distance a c.

Then say: The square of a c divided by a d equals a b, the distance required.

EXAMPLE.—Given a c 26 ft., and a d $13\frac{1}{2}$ ft.; required the distance a b.

STATEMENT.—$26^2 \div 13\frac{1}{2} = 50$ ft., distance a b.

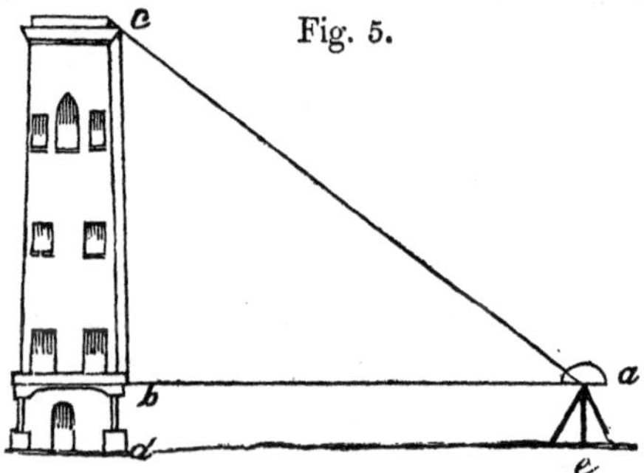

Fig. 5.

We wish to find the height of a tower, or building, as d c.

Set your instrument any convenient distance, e, neither too great nor too small, in comparison to the altitude d c, and measure the angle b a c, and measuse the distance a b or e d; you then have, in the right-angle triangle, one side given, and the angle b a c.

RULE 18.

Multiply the nat. tangent of the angle b a c by the distance a b; will equal b c.

NOTE.—The point at *b* can be observed, and afterwards th distance *b d* can be measured, which, added to *b c*, will determine the distance *c d*.

In finding the height of an object, let the observed angle be as near 45° as possible; for then a small error committed in taking it, makes the least error in the computed height of the object; because, if the observed angle, as at *a*, equals 45°, the distance *a b* will equal *b c*.

It very seldom occurs, in the construction of a railroad, to measure verticle heights with an instrument.

In running levels, if the top of a hill is found inaccessible to find the elevation with a leveling instrument, we have to resort to the examples given by triangulation.

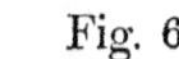

Fig. 6

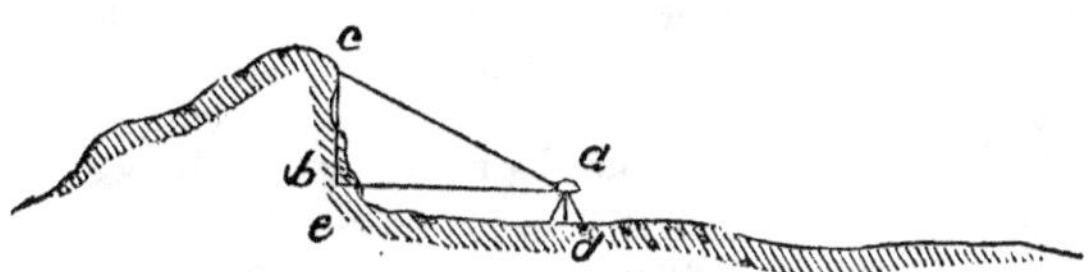

When it is necessary to determine the elevation of a hill as in Fig. 6, the elevation at *d*, is found with the leveling instrument, as will be explained in the Art of Leveling.

EXAMPLE.— Set your instrument over *d*, and measure the angle *b a c*, and distance *a b*; you then have for the triangle *b a c*, the angle and one leg given, to find the other leg, *b c*, as in Fig. 5; which, added to the height of your instrument, *a* from *d*, will equal *e c*; and added to your elevation *d*, will equal the elevation *c*.

THE ART OF LEVELING.

The first thing necessary in leveling, is to have the requisite instruments in adjustment.

To Adjust a Level.

In the common Y level there are three adjustments.

1st ADJUSTMENT.—Place the instrument in a firm position, and unclamp the Y's; place the horizontal hairs on some distant object, and revolve the telescope half around; if the hair intersects the point first observed, the instrument is in adjustment, thus far; if not, move the hairs half way distant, between the two points of intersection, by means of the screws on the telescope, generally marked "Hairs," and by revolving the telescope, the hairs will intersect our given point. The vertical hairs can be adjusted the same way.

2d ADJUSTMENT.—With the instrument firm, as before.—Fasten the telescope over the leveling screws, and level it exact; then take the telescope out of the Y's and reverse it; if the bubble is level, this adjustment is correct; if not, divide the difference of the bubble (one-half,) by means of the screws under the bubble, and level the remainder by means of the leveling screws. This process for the second adjustment hardly ever proves correct the first time; therefore, repeat the above, until the telescope, when revolved in the Y's, on every screw, the bubble will be level.

3d ADJUSTMENT.—After the above adjustments, fasten the telescope on the Y's, by means of pins generally used; place your telescope over the leveling screws, and bring the bubble to a level, and repeat it on all the screws, so as to get the telescope as level as possible before commencing the adjustment; then placing the telescope over any two of the leveling screws, and level the bubble; reverse the telescope half way on the pivot, or, as near as possible over the same screws; if the bubble is level, the adjustment is correct, if not, move the bubble half way, by means of the screws under the leg of the Y, and level the remainder by means of the leveling screws. By continuing this process on all the screws, the adjustment can be perfected.

NOTE.—The last adjustment is immaterial, only in saving time and trouble when using. The difference (if there is any,) is so comparatively small, that it is not observable.

The third adjustment never will remain in adjustment on most of levels, so that no trouble need be borrowed when it is found that your level will not reverse correctly. The adjustment of the level now being complete, we will proceed to its use.

In preliminary surveys, or location of railroads,

levels have to be run (as it is termed) to ascertain the exact surface of the ground, in order to establish grades. In commencing levels, an elevation is established upon a given point. This point is generally made by cutting on the root of a tree and is termed a "Bench." These benches are established on the entire length of the line, perhaps one-half to three-quarters of a mile apart, for reference points.

This elevation is generally estimated above, so as to reach the lowest point of the surface of the ground that should occur in your levels. For instance: The lowest point of ground we guess to be 50 ft. below the first established bench, and for safety would call the bench elevation 60. We set up our level firm in the ground, not to exceed 400 ft. from the bench, and near the line we wish to run the levels over, and take what is termed a back sight (marked B. S.) on the bench, by holding the staff, or leveling rod, on the bench, and moving the target of the rod to its intersection with the horizontal hairs in the telescope; what the rod would read at this intersection, would show that our instrument would be that number of feet and parts above the bench. For instance: Suppose the rod read $3\frac{416}{1000}$ ft., therefore, the elevation of the instrument would be 63·416.

When we have the height of the instrument given, it shows very plainly that if you take a sight at any given point, the elevation would be as much less as the rod would read. For instance: Suppose the rod at any point, or station, (as stations of 100 ft. are used in the location of railroad lines,) should read $10\frac{198}{1000}$, which are termed fore, or intermediate sight, (marked F. S.) the elevation of that point taken would be 53·218; or, 63·416 — 10·198 = 53·218; consequently, these fore sights should be subtracted from the height of the instrument, or elevation of the instrument, to give the elevation of stations.* In order to keep up the same corresponding elevations, on the entire length of the line, we change our instrument on some substantial point, as a peg or stone, by holding the rod upon the peg; being a fore sight, we subtract it from the height of the instrument, which gives the elevation of the peg, and is the same as a bench. Suppose the rod reads on the peg 8·747, and instrument is 63·416; 63·416 — 8·747 = 54·669, elevation of peg. †

We have the elevation of the peg, and can move our instrument further on, and set up our instrument firm in the ground, as before, and take a "back sight" on the peg, by holding the rod upon the peg, and notice the reading as before. Suppose the rod to read 1·201; it shows that our instrument is 1 foot and $\frac{201}{1000}$ above the peg; consequently, if we add it to the elevation of the instrument, thus: peg = 54·669, rod reads 1.201 + 54·669 = 55·870, height of instrument, and proceed as before, taking intermediate sights, subtracting every intermediate from the

---

* These fore sights are more properly termed intermediate sights (marked I. S.) which we will hereafter term them, and fore sights will be termed as at changes of the instrument, after running the level of intermediates, not to exceed 400 ft. on either side of your instrument.

† In practice, whether you change your instrument upon a peg, stone, stump, or anything suitable, it is termed a peg.

height of the instrument. It is not necessary to carry out the reading of the rod of the decimals, (when taking intermediates,) farther than hundredths, as 2·27, 4·15, 8·11, &c., and in most cases tenths is far enough, as 2·2, 4·1, 8·1, &c., &c.

Intermediates, or plus stations, should be taken when the ground varies to any amount; discretion on your own part must govern that. It is evident that if you were running levels over an uneven ground, as Fig. 1, and should take the elevation at station 1 and station 2, that you lose, or there would be a loss in the estimate of the quantities, or if for the purpose of establishing grades, a correct line could not be drawn, as a man's discretion is governed by the correctness of the profile, or levels taken, therefore, a level should be taken at *a*, also at *b c d*, and the plus station noted in the book of levels, then you have the correct shape of the ground.

Fig. 1.

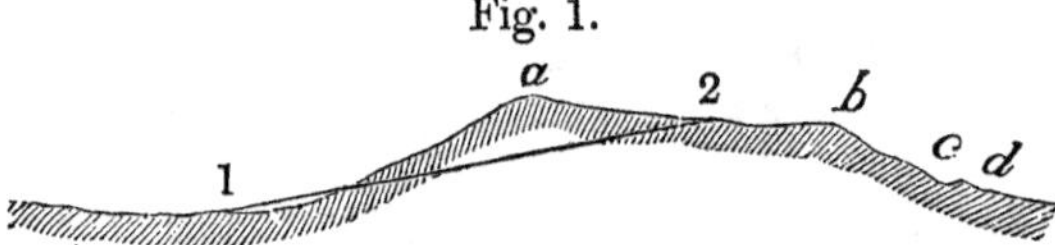

To explain the foregoing more intelligibly, we will refer to Figs. 2, 3, and 4.

Our established bench elevation 60, is at *A*, on the root of a stump; our line to run is in the direction of *d*, consequently, we would set our level at *c*, not to exceed 400 ft. from the bench *A*; in directing our level at the target *f*, we find it reads 3·416; then our level would be that number of feet and parts above the point *a*; bench elevation is 60, elevation of instrument would be 60 + 3·416 = 63·416; we will take the elevation of the ground at the stations 1, 2, 3, 4, 5, 6, &c., reading the nearest tenth of a foot, on the rod. Station 1 reads 3·5, therefore, 63·416 — 3·5 = 59·916, or 59·9 elevation of station 1. Station 2 reads 3·3, then 63·4 — 3·3 = 60·1. Station 3 reads 3·6, then 63·4 — 3·6 = 59·8. Station 4 reads 3·7, then 63·4 — 3·7 = 59·7. Station 5 reads 3·6, then 63·4 — 3·6 = 59·8. Station 6 reads 3·9, then 63·4 — 3.9 = 59·5. Station 7 reads 4·8, then 63·4 — 4·8 = 58·6. Station 8 reads 5·8, then 63·4 — 5·8 = 57·6. Station 9 reads 8.4, then 63·4 — 8·4 = 55.0.

Fig. 2.

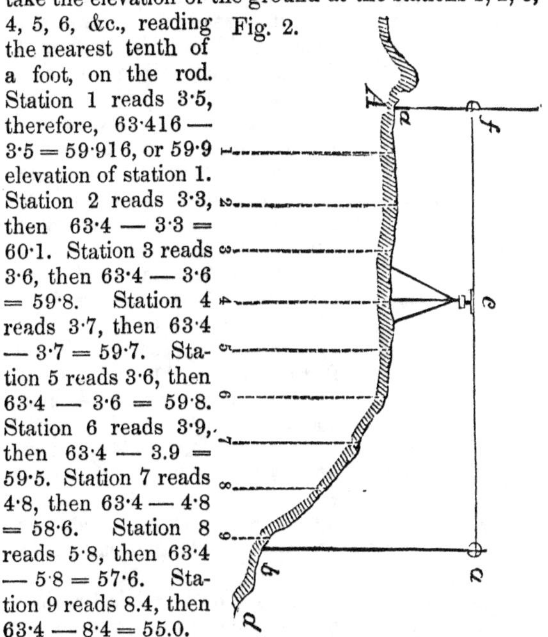

We will change our instrument at station 9, for convenience. We drive in a peg, firm into the ground, at or near the station; the rod reads 8·747, then 63·416 — 8·747 = 54·669, elevation of the peg at *b*. This being a secured point, we move our instrument, as in Fig. 3, to *f*. After firmly setting our instrument, we take a back sight on the peg *b*. Suppose it to read 1·201, the instrument would be that number of feet and parts above the peg. Elevation of the peg we found to be 54·669; the elevation of the instrument would be 54·669 + 1·201 = 55·870. We now have the elevation of the instrument again. For intermediates, Station 10 reads 2·2, then 55·8 — 2·2 = 53·6. Station 11 reads 3·1, then 55·8 — 3·1 = 52·7. Here we should take an elevation at *a*. Suppose it to be 50 ft. from station 11, as found by measurement; then the plus station would be noted 11 + 50. Station 11 + 50 reads 3·4, then 55·8 — 3·4 = 52·4. Station 12 reads 3.1, then 55·8 — 3·1 = 52·7. Station 13 reads 2·6, then 55·8 — 2·6 = 53·2. Station 14 reads 2·1, then 55·8 — 2·1 = 53·7. Station 15 reads 1·8, then 55·8 — 1·8 = 54·0. Station 16 reads 2.8, then 55·8 — 2·8 = 53·0. Station 17 reads 3·2, then 55·8 — 3·2 = 52·6.

Fig. 3.

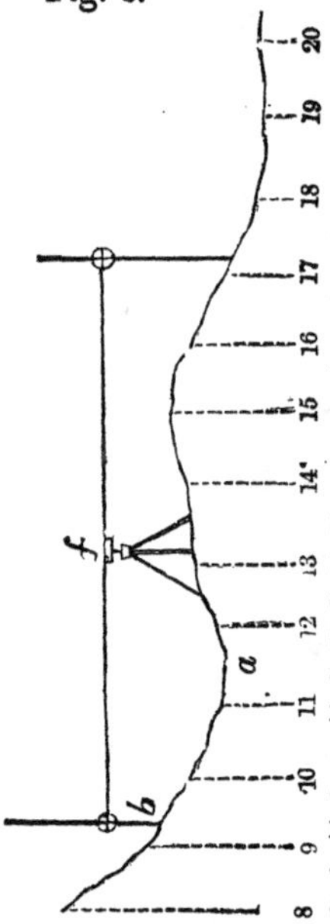

We will change our instrument by driving a peg

at or near station 17, and finding its elevation.* Suppose the F. S. on the peg reads 3·775, then height instrument 55·870 — 3·775 = 52·095 = elevation of peg; we then move our instrument to *f*, (Fig. 4) and take a B. S. on the peg. Suppose it to read 2·000; elevation peg 52·095 + 2·000 = 54.095 = height of instrument. Then proceed as before.

Fig. 4.

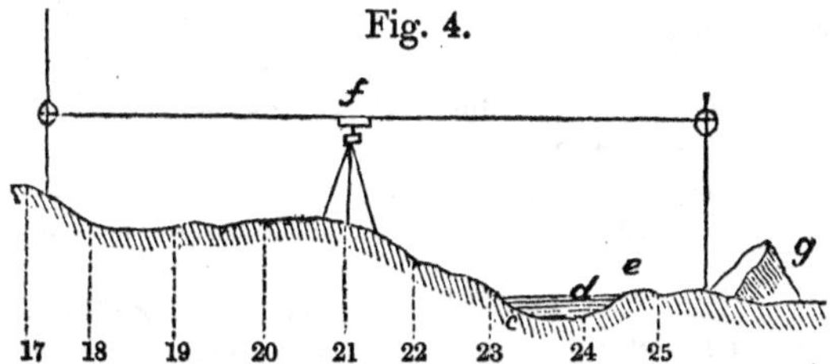

Station 18 reads 3·0, then 54·1 — 3·0 = 51·1.
Station 19 reads 2·9, then 54·1 — 2·9 = 51.2.
Station 20 reads 2·7, then 54·1 — 2·7 = 51·4.
Station 21 reads 2·5, then 54·1 — 2·5 = 51·6.
Station 22 reads 3·6, then 54·1 — 3·6 = 50·5.
Station 23 reads 4·4, then 54·1 — 4·4 = 49·7.

Here at station 23 we have come to a stream, and it is quite necessary to get its shape, width, and character. We have got the elevation of the top of the slope of the river at station 23; we take an elevation at *c*, at the bottom of the slope of the creek, and notice its plus distance from station 23. Suppose it to be 25 ft.; then 23 + 25 would be the bottom

* When changing, the elevations should be exact.

slope of the stream. We also get the elevation of the bottom of the slope on the opposite side *d*, and notice the plus station. Suppose it to be 85 ft. from station 23, then it would be 23 + 85, bottom slope of stream; and station 24 would also be taken, as we very seldom miss stations, wherever they may occur We have the top slope 23, and two bottom slopes 23 + 25, and 23 + 85, and to complete the levels of the creek we want the top slope of opposite side. Say it is 75 ft. from station 24; then elevation at 24 + 75 gives the shape of the river's banks and bottom. Also to govern the masonry, or bridging, over this stream, the elevation of high water mark is taken. This will be readily found by inquiry, if no signs on the banks are visible.

Other very important things have to be observed and placed under remarks, which your own discretion must lead you.

We will establish a bench before proceeding farther. Say, for instance, we cut a notch on the root of the stump *g*. Suppose our rod, or F. S., reads 4·005; height of instrument 54·095 — 4·005 = 50·090, elevation of the bench. The elevation of the benches are generally marked on the stump with either red chalk or paint. We secure this elevation at *g*, and along the line at intervals, to avoid the trouble of going back to the first established bench, *A*, Fig. 2, on the line. For instance: Suppose it is required to find the elevation of station 25, or wished to stake out a bridge, or stone culvert, in the stream at some future day. We, instead of running from the first established bench, take the established bench *g*, by setting our instrument say at *f*, the most convenient point, and do the work required.

We proceed with our levels as before, and establish benches, or elevations, not to exceed three-quarters of a mile apart.

It might be necessary to state, for accuracy in running levels, that the rod should be plumb, or point to the centre of the earth; also the bubble of the level should always be level. Accuracy depends both upon the leveler and rodman.

I will here give the manner of keeping a field-book for running levels.

NOTE.—In running levels, always add the back sights, which will give the elevation of the instrument, and subtract the fore or intermediate sights, will give the elevation of the peg or bench.

MANNER OF KEEPING FIELD-BOOK.

NOTES FROM FORMER EXAMPLES.

Sta'n.	B. S.	F. S.	I.S.	H. Inst.	Elev' n.	Remarks.
B'ch	3·416			63·416	60·000	On root of stump near stat'n 1.
1			3·5		59·9	
2			3·3		60·1	
3			3 6		59 8	
4			3·7		59·7	
5			3·6		59·8	
6			3·9		59·5	
7			4·8		58·6	
8			5·8		57·6	
9			8·4		55.0	
Peg.		8·747			54·669	Elevat'n of peg at B. (Fig. 2.)
10	1·201		2·2	55·870	53·6	
11			3·1		52·7	
+50			3·4		52·4	

Stat'n.	B. S.	F. S.	I. S.	H. Inst.	Elevat'n.	Remarks.
12			3·1	63·416	52.7	
13			2·6		53·2	
14			2·1		53·7	
15			1·8		54·0	
16			2·8		53·0	
17			3·2		52·6	
Peg.		3·775			52·095	
18	2·000		3·0	54·095	51·1	
19			2·9		51·2	
20			2·7		51·4	
21			2·5		51·6	
22			3·6		50·5	
23			4·4		49·7	Top of bank of stream.
+25			5·2		48·9	Bottom of stream.
+85			5·3		48·8	Bot. of stream, opposite side.
24			5·0		49·1	
+75			4·4		49·7	Top slope of opposite side.
			4·6		49·5	High water mark.
B'ch		4·005			50·090	On stump by station 26.

In leaving the work at night, benches should be made, so that when you choose at any time to go to work, you have a convenient place to commence, and accurate.

After the levels are run the work is plotted and grades are established.

Grades vary in their ascent according to circumstances, and are governed by the discretion of the engineer in charge. They intend however to equalize the excavation and embankment as near as possible. Grades sometimes can be improved, and are governed by the contracts taken to grade the road.

Grades have elevations as well as the surface of the ground. We will assume the grade at Station

1 to be 58·000 and descent is $0 \cdot \frac{22}{100}$ per 100 feet, that it equalizes the excavation and embankment. We then have elevation at

Station 1 = 58·000	Station 13 = 55·360
" 2 = 57·780	" 14 = 55·140
" 3 = 57·560	" 15 = 54·920
" 4 = 57·340	" 16 = 54·700
" 5 = 57·120	" 17 = 54·480
" 6 = 56·900	" 18 = 54·260
" 7 = 56·680	" 19 = 54·040
" 8 = 56·460	" 20 = 53·820
" 9 = 56·240	" 21 = 53·600
" 10 = 56·020	" 22 = 53·380
" 11 = 55·800	" 23 = 53·160
" 12 = 55·580	" 24 = 52·940

With the elevation of the surface of ground, and the elevation of grade, their difference would be the cut or fill. For instance

—— Elevation of the surface Station 1 = 59·9
grade " . 1 = 58·0
Difference = 1·9

The difference shows the cutting 1·9 as the surface elevation is the greatest, and at Station 11 Surface elevation = 52·7, Grade elevation 55·8, difference 3·1, fill, as the surface elevation is the least. When the estimates in cubic yards have to be made, (as they are generally made approximately from the profile) the cutting and filling is easily ascertained by calculating the grade for every 100 feet, and taking the difference of elevation as just shown.

When a line is located, the grades are then established, and construction commences. We then have different field books, termed Grade Book, Cross Section Book and Monthly Estimate Book.

MANNER OF KEEPING GRADE BOOK.

Sta'n.	Grade.	Cut.	Fill.	Elv'n	Remarks.
1	58·000	1·9		59 9	
2	57·780	2·3		60·1	
3	57·560	2·2		59·8	
4	57.340	2·4		59·7	
5	57·120	2·7		59·8	
6	56.900	2·6		59·5	
7	56·680	1·9		58·6	
8	56·460	1·2		57·6	
9	56·240		1·2	55·0	
10	56·020		2·4	53·6	
11	55·800		3·1	52·7	
+50	55·690		3·3	52·4	
12	55 580		2·8	52·7	
13	55·360		2·1	53·2	
14	55·140		1·4	53·7	
15	54 920		0·9	54·0	
16	54 700		1·7	53 0	
17	54·480		1·8	52·6	
18	54·260		3·1	51·1	
19	54·040		2·8	51·2	
20	53.820		2 4	51·4	
21	53·600		2·0	51·6	
22	53·380		2·9	50·5	
23	53·160		3·4	49·7	Top of bank of stream.
+25	53·105		4·2	48·9	Bottom of stream.
+85	52·973		4·1	48·8	Bottom of stream opposite side.
24	52·940		3·8	49·1	
+75	52·775		3·0	49·7	Top of slope opposite side.

The benches are all entered in the back part of the grade book for reference when necessary. In the grade book we have the cuts and fills worked out, but in the staking out of the work, these cuts and fills are merely used for tests. Engineers generally have in all their work test points for reference, which all should have, to avoid the many mistakes that will occur.

In staking out work the grade point or where the excavation and embankment commences, is always found with its distance from the station joining, and a stake put in to guide the contractors in commencing their work.

The grade points are generally marked in the cross section book.*

Cross sections are cuts and fills taken at right angles to the line of the road, any distance from the center line that should seem necessary by the engineer.

Cross sections are taken with the level, but more plus stations are taken, than in running levels, as the discretion of the engineer may direct him. Cross sections are taken to get as near as possible the amount of cubic yards excavation and embankment, as contracts are taken of the work, to complete at a given price per cubic yard. On very level ground, cross sections are taken at every station. On very uneven ground cross sections are taken as often as your judgment dictates.

Before proceeding further, we will explain the manner of getting the cuts and fills with a level in the field.

---

*This cross section book is more properly termed Original Cross Section Book, as there also is the Final Cross Section Book.

With the grade book we have the elevation of grade. We go into the field, say at station 21, and set up our level, and take a B. S. on the bench *g*, (Fig. 4,) and find the elevation of our instrument. Suppose the elevation of the bench to be 50.090, and rod reads 5·112, then 50·090 + 5·112 = 55·202, elevation of instrument. Suppose we wish to commence at station 19, with our cross sections. We would refer to the grade book at station 19 and find the elevation of grade at that point = 54·040. We now have the elevation of our instrument and the elevation of the grade at station 19. Now if we subtract the elevation of the grade from the elevation of the instrument, the difference shows that our instrument is that number of feet and parts above the grade line at station 19. For instance:

Elevation of instrument =	55·202
grade =	54·040
Difference of elevation =	1·162

Then our instrument at station 19 is 1.162 feet above the grade line. We will get the cut or fill at station 19. Suppose our rod to read, at station 19, 4·0, (the nearest tenth of a foot,) it would show our instrument to be four feet above the surface of the ground, and if our instrument is 1·162, (or nearest tenth,) 1·2 above the grade line, and 4·0 above the surface, we see that if we take their differences, it will give the fill or cut at station 19. Then—

Elevation of instrument above surface of ground =	4·0
grade =	1·2
Difference =	2·8

2·8 is then the fill at station 19. We then take our cross section at station 19, subtracting the height of instrument above the grade line, 1·2, from the height of the instrument above the surface, will equal the cut or fill at station 19, any distance at right angles from the station or center line.

We see that our instrument at station 20 (by reference to a profile) would not be 1·162 above the grade line, but would be more as the grade descends. We have found that the grade descends $0\cdot\frac{22}{100}$ per 100 feet. Then if we add 0·22 to the instrument above the grade line at station 19, will equal its height above the grade line at station 20. Then elevation of instrument at station 19 = 1·162; descends 0·22 per 100 feet, or station, then 0·22 + 1·162 = 1·382; height above station 20 = 1·382.

We will suppose the rod reads at station 20, 4·0, the same as station 19, then 4·0 — 1·4 (nearest tenth of 1·38) = 2·6 fill, or 2 feet and 6 tenth is the difference between the grade line and the surface of the ground at station 20.

We can continue this process to all the stations, until it becomes necessary to change the instrument. When we change, the elevation of the instrument we have preserved = 55·202, and change on a peg, as in running levels, subtracting the sight on the peg for its elevation, and after the instrument is set up again, add the sight to the elevation of the peg, and you have the height of your instrument again, and proceed as you commenced.

Sometimes we find our instrument below the grade line. We will suppose that the rod reads on the bench *g*, (Fig. 4,) 2·950; this added to the bench would give the elevation of the instrument; bench = 50·090 + 2·950 = 53·040, height of instrument. Grade at station 19 = 54·040. Now we have the elevation of the instrument, less than the elevation of the grade—

Thus, elevation of instrument = 53·040
" grade at stat. 19 = 54·040
Difference of elevation = 1·000

Shows that our instrument is 1 foot below the grade line. Now if our instrument is below grade, the sights on the surface must be added to the elevation of the instrument, to give the difference or distances of the grade line to the surface. Suppose the rod, or sight, at station 19, reads 1·8. We would see that the surface of ground was 1 foot and 8 tenths below the instrument, and the instrument is 1 foot below the grade line, therefore the distance from the grade line to the surface of the ground, would be their sum. Thus,

Elevation of instrument below grade = 1·00
" " above surface = 1·8
Difference of surface and grade line = 2·8

2·8 would be the fill at station 19.

When you come to grade, the elevation of the instrument above the grade line, and its elevation above the surface of the ground, will be equal.

For example, suppose the instrument be 1·2 above the grade line, and surface of the ground 1·2, the difference is 00, and is the point of grade.

If the elevation of the surface and elevation of grade are equal at any point, that part of the surface of the ground is grade, or the point where the excavation and embankment commences. Thus,

Elevation of grade = 54·040
Elevation of surface of ground = 54·040
Difference = 0·00

In taking cross sections, the elevation of the instrument should be preserved for changing — and to be correct, you should keep a table of both your elevation of peg and instrument and height of instrument above the grade line at each and every station. To show more plainly, we will make a sketch of the work already done:

Bench =	50·090
B. S. =	5·112
H. Inst. =	55·202
Change Inst., F. S. =	2·211
Peg =	52·991
B. S. =	0·049
H. Inst. =	53·040

Also,— Grade at stat. 19 =	54·040
H. Inst. =	55·202
H. Inst. at stat. 19, above grade line =	1·162 +
Grade descends 0·22 per 100 feet =	0·22
H. Inst. at stat. 20, above grade line =	1·382 +
	0·22
H. Inst. at stat. 21, above grade line =	1·602 +
	0·22
H. Inst. at sta. 22, above grade line =	1·822 +

&c., &c. Change instrument.

We have the height of instrument again, 53·040, and will commence at station 19 again to show the manner of keeping notes with the instrument below grade:

Grade at stat. 19	=	54·040
H. Inst.	=	53·040
H. Inst. at sta. 19	=	1·000—
" " 20	=	0·780—
" " 21	=	0·560—
" " 22	=	0·340—

The sign of plus and minus can be annexed to the above, to show that the instrument is above or below grade line. Thus, + = above, — below.

In running from station 19 down the grade, we notice that we add the descent, per 100 feet, to the height of the instrument when plus, and when minus we subtract, because when the instrument is above, the further we run on the descent the greater the height of the instrument from the grade line, as much greater as the descent per 100 feet; and when the instrument is minus, we subtract the descent, because we near the grade line every 100 feet, as the descent of the grade per 100 feet.

Sometimes, in staking out work, it is not necessary

Fig. 5.

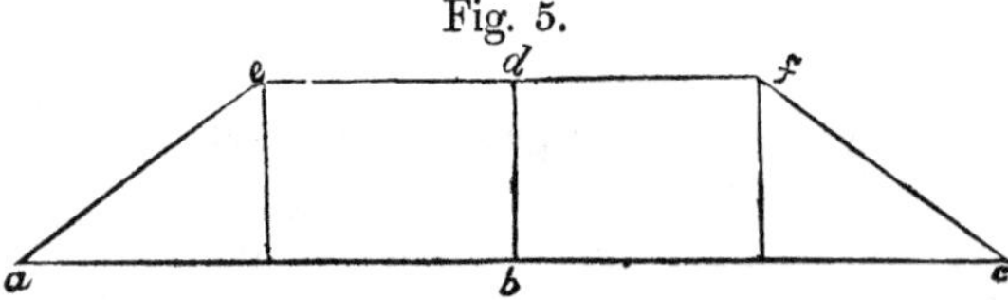

to cross section further than where the slope stakes would occur, as in Fig. 5.

The above shows a cross section, *a b c*, with the slopes *a e f c*, determined, and slope stakes *a c* put in.

The slope of a railroad is governed by the material of which it is composed.

The excavations are governed the same as the slopes of embankments; the width at the top is governed by the width of track or material. We will assume it to be 14 feet.

In making a cross section of the above, we will assume slopes of $1\frac{1}{2}$ feet horizontal to one foot vertical, as the required slopes. Suppose we take a level at *b*, (surface of the ground and center of the road,) and find the fill to be four feet. We mark the stake *b* with red chalk, "4 feet fill," (this is to guide the contractor,) and measure out from *b*, on either side towards *a*, one-half the width of the road bed, 7 feet, and take the cutting or filling. From this filling, we can judge of the point for a slope stake *a*, if the ground does not vary too much; if it should, we keep trying until the exact point for the slope stake is found. For example, we take the level at *a*, and find that there is $3\cdot\frac{3}{10}$ feet filling, and, by calculation, slopes $1\frac{1}{2}$ to 1, we measure from the center 7 feet, and base of slope $5\cdot\frac{7}{10}$, making in all to measure from the center line, (as one end of the tape or chain should always be kept at the center stake to avoid confusion in changing from one side to the other,) would equal $12\cdot\frac{7}{10}$. We take another sight if it varies much from the point in which we took our sight, and make the same calculations again. It may vary

2 or 3 tenths in the distance from the point last found, but if the ground is level, or nearly so, the required distance can be measured for the slope stake. You go through the same process on the opposite side for the slope stake *c*. We now see that if the contractor fills in 4 feet at *b* to *d*, and 7 feet from *d* to *e*, and 7 feet from *d* to *f*, and carries the dirt out to the slope stakes *a* and *c*, that the slope of the cross sections would have 1½ feet horizontal to 1 foot vertical. Slopes more general in use are 1½ feet horizontal to 1 vertical; 2 feet horizontal to 1 vertical; 1 horizontal to 1 vertical. Calculations for slopes will be found in the following rules:

Slopes 1½ to 1.—With the filling or cutting given, to find the length of the base of the slope.

RULE 1.

One half the cut or fill added to the cut or fill where it is taken.

EXAMPLE.—Given the filling 3·8, to find the base of the slope.

STATEMENT.—½ of 3·8 = 1·9 + 3·8 = 5·7.

Slope 2 to 1.—With the filling or cutting given as before, to find the base of the slopes.

RULE 2.

Multiply the cutting or filling by 2.

EXAMPLE.—Given the filling 3·8 to find the base of the slope = 3·8 × 2 = 7.6.

Slope 1 to 1—

RULE 8.

The base is equal to the filling or cutting.

EXAMPLE.—Given the filling or cutting 3·8; then the base would equal 3·8.

CROSS SECTIONS.

The number of cross sections to be taken will be as the engineer's judgment governs him, but we will give a few examples necessary to correctness in side hill ground entering from a cut to a fill. The following figures 6, 7 and 8 will represent the cross sections.

In leaving the cut (Fig. 6) a cross section should be taken where the grade point occurs at the bottom slope as at *a*, also a cross section should be taken (Fig. 7) where the grade point occurs at the center B; also where the grade point occurs on the opposite side *c*. (Fig. 8.)

Cross sections taken in this manner, give the shape of the ground as near as can be got at. We also have them in plane figures. The area of Fig. 6 to correspond with the area B *c d* (Fig. 7) and Pyd *a B e*, also the Pyd *B c d*, the area *a B e* to correspond with the area *a b c* (Fig. 8). The quantities in this manner are got as correct as is possible. Cross sections around curves should not exceed 50 ft. apart when the average form is used for calculating quantities.

Henck gives a formula for estimating the quantities in curves, but it has been adopted only by a very few engineers, if any. My mode is most in practice by eminent engineers at present, but there is no doubt that the manner in which Henck and also Trautwine do their work will come into general practice, as the immense quantities of earth that is estimated by the average form would be very materially diminished.

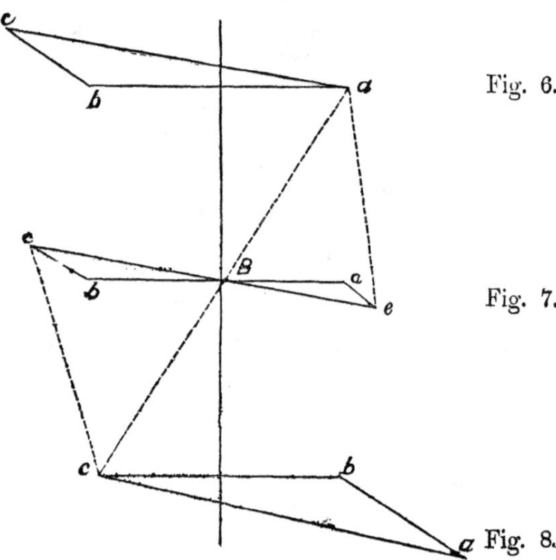

Fig. 6.

Fig. 7.

Fig. 8.

BORROWING PITS

Consist of borrowed earth from a hill or side of a mountain, when the earth in the cut is not sufficient for the embankments.

It very often occurs where earth has to be borrowed to finish up embankments; sometimes it is taken from ditches; sometimes knolls are cut off, and sometimes it is taken out of the side of hills and mountains. This amount taken out has to be ascertained in cubic yards.

The Manner of Measuring Borrowing Pits, to ascertain the Quantities taken out.

The object in the first place, is to make original surveys or cross sections; and secondly to make cross sections, (after the earth is taken out) to correspond with the original cross sections. In order to do this, points must be established in the original survey, that can be referred to when you wish to make the second survey. Sometimes it has to be measured monthly, to make monthly estimates.

In order to get correct measurements we will establish a line, (called a Base Line) along the base of the hill* between two points, that will cover the length of the area. For security, the points we will establish upon the roots of some stumps, and in a direct line between these two points, put in stations and plus stations as often as is considered necessary. From these stations, at right angles to the established

* This is governed altogether by the engineer's judgment.

line, measure with the tape or chain, and where it is necessary take sights with the level, and ascertain the elevation, or cut or fill above a certain given point. Continue on this measurement and elevations back as far as will cover the area to be excavated in that direction; continue the same process at every station and plus station. The notes will be entered in the original cross section book.

If it is necessary to make a monthly estimate, the base line is found, and stations and plus stations are put in as before, and measurements are commenced as before. This is entered into the monthly cross section book.

When the work is completed, a final measurement has to be made, by finding the base line, and proceed as in making the original survey.

These elevations are taken with the level. A base line of levels is established for the pit, and when sights are taken, the notes show the elevation of the ground, either above or below the established base line of levels. When the last survey is made, where the earth has been taken out, the difference of elevation at their corresponding points would show the depth of earth taken at these points.

When those measurements are made, they are plotted upon paper, which will show the area of the cross section.

For example, suppose *a b* (Fig. 9) to be the base line of levels.

The original levels commence at the base line of stations, 1 foot 5 tenths above the base line *a b*, and at the top of the hill, or the extent of excavation, the elevation is 13 feet 4 tenths above the base line. When the second measurement is made, we find at the station to be the same as the original, 1 foot 5

Fig. 9.

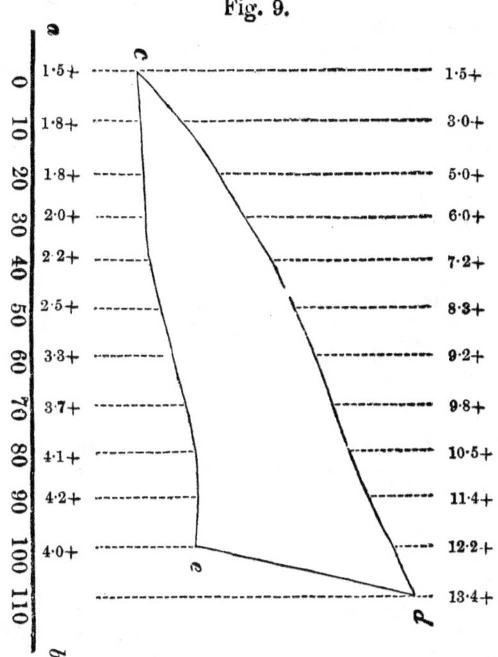

tenths, and find the last point to be 4 feet above the base line. The excavation we find at the top to be

110 feet from the point at *c*, and at the bottom, *e*, or second survey, 100; this will make the slope *d e*, and complete the area *c d e*.

The depth of excavation can be ascertained by the difference of the elevations.

For instance, the depth of cutting 80 feet from the station *c* equals the difference of the elevations 4·1 and 10·5 = 6·4; the cutting then at 80 feet distance is 6 feet 4 tenths. The remainder of the area is ascertained in the same manner. In the notes the sign of + is annexed to the sights when above the base line of levels, and — when below.

If in two corresponding sights or elevations, one should be below the base line of levels and the other above, the depth of excavations would equal their sum.

If in two corresponding sights or elevations, both should be below the base line of levels, their difference would equal the depth of excavation; if both should be above, their difference would equal the depth of excavations.

NOTE.— All horizontal measurements, with a tape or chain, either in measuring for cross sections or land, should be measured perpendicular to the center of the earth.

The area of the cross sections (Fig. 9) is easily determined, as the original elevations and final elevations correspond in their distance from the station *c*. It is not always that cross sections can be taken with equal distances (10, 20, 30 feet, &c.) from the station, as the ground may be very irregular both in the original and final measurement. The area of the section is ascertained by calculating the area of the original measurement to the base line of levels, and the area of the final measurement to the base line of levels; the difference of these areas would be the area of the section.

Fig. 10.

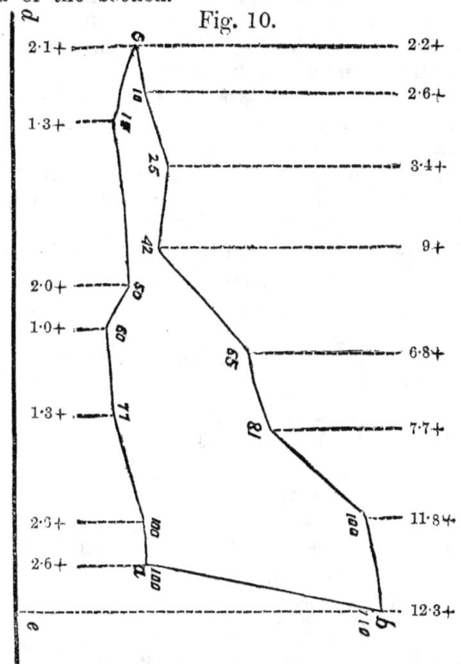

For example, to ascertain the area of section *a b c*, (Fig. 10.)

Instead of the distances 10, 20, 30, &c., from the station at *c*, we have been obliged (on account of the unevenness of ground) to measure the distances 10, 25, 42, 65, 81, 100 and 120 from the station *c*, and in the final the distances 15, 50, 60, 77, 100 and 110. In this case the depth of excavation cannot be ascertained, but the area is easily found.

We wish to get the area of the section *a b c*. By computing the area *b c d e*, and area *b e d c a*, the difference of their areas would equal the area *a b c*.

The manner of keeping original cross section book and final cross section book, is similar. For example, we will arrange Fig. 9 and Fig. 10, supposing that Fig. 9 = stations 1, and Fig. 10 = stations 1 + 20.

ORIGINAL CROSS SECTIONS.

(Base Line or Center.)

Stations 1.

1·5+	3·0+	5·0+	6·0+	7·2+	8·3+	9·2+	9·8+	10·5+	11·4+	12·2+	13·4+
	10	20	30	40	50	60	70	80	90	100	110

Station 1 + 20.

2·2+	2·6+	3·4+	2·9+	6·8+	7·7+	11·8+	12·3+
	10	25	42	65	81	100	120

FINAL CROSS SECTIONS.

Station 1.

1·5+	1·8+	1·8+	2·1+	2·2+	2·5+	3·3+	3·7+	4·1+	4·2+	4·0+
	10	20	30	40	50	60	70	80	90	100

Station 1 + 20.

2·2+	1·3+	2·0+	1·0+	1·3+	2·6+	2·6+
	15	50	60	77	100	110

When the above is plotted, it forms the figures 9 and 10.

The number of cubic yards contained between

these two figures (according to the present mode of calculation) is equal to one-half the sum of their areas, multiplied by the distance they are apart, (20 feet,) and the quotient divided by 27.

MENSURATION OF SURFACES.

To find the area of a right-angled triangle.

RULE 1.

Multiply one-half the base by the perpendicular height, equal the area.

To find the area of a triangle.

RULE 2.

Multiply the base by the perpendicular let fall on to it from the opposite angle, and one-half the product equals the area.

To find the area of a triangle by its sides.

RULE 3.

From half the sum of the three sides subtract each side separately; then multiply the half sum and the three remainders continually together, and the square root of the product will equal the area.

To find the area of a rectangle or a square.

RULE 4.

Multiply the perpendicular height by the length, equal the area.

To find the area of a rhombus or a rhomboid.

RULE 5.

Multiply the length by the perpendicular distance let fall from its sides, equals the area.

To find the area of a trapezoid.

RULE 6.

Multiply one-half the sum of the parallel sides by the perpendicular, equals area.

To find the area of a trapezium.

RULE 7.

Multiply the diagonal by the sum of the two perpendiculars falling upon it from the opposite angles, and half the product equals area.

To find the area of a regular polygon.

RULE 8.

Multiply one of its sides into half its perpendicular distance from the center, and this product into the number of sides, equals its area.

To find the area of an irregular polygon.

RULE 9.

Draw diagonals to divide the figure into trapeziums and triangles; find the area of each separately, and the sum of the whole equals area.

To find the area of a long irregular figure, bounded on one side by a straight line, (Figs. 9 and 10 on borrowing pits.)

RULE 10.

Multiply one-half the sum of each succeeding height by their distance apart, the product will be the area between the two heights, the sum of all the areas will equal the area of the figure.

To find the area of a circle when the diameter and circumference are both known.

RULE 11.

Multiply the square of the diameter by ·7854, or the square of the circumference by ·07958; or multiply the circumference by the diameter, and divide the product by 4, will equal area; or one-half the circumference by one-half the diameter.

To find the area of a sector of a circle.

RULE 12.

Multiply the length of the arch by the radius of the circle, and half the product will equal the area, (nearly.)

To find the area of a segment of a circle.

RULE 13.

Multiply the versed sine by the decimal ·626; to the square of the product add the square of half the chord; multiply twice the square root of the sum by two-thirds of the versed sine, will equal area.

To find the area of an ellipsis.

RULE 14.

Multiply the transverse or longer diameter by the conjugate or shorter diameter, and by ·7854,* will equal area.

To find the area of a circular ring or space included between two concentric circles.

RULE 15.

Add the inside and outside diameters together, multiply the sum by their differences, and by ·7854, will equal area.

To find the area of a parabola or its segment.

RULE 16.

Multiply the base by the perpendicular height, and two-thirds of the product equals area.

MENSURATION OF SOLIDS.

To find the solid contents of a cylinder.

RULE 1.

Multiply the area of the base by the height of the cylinder, and the product is the solid contents.

To find the solid contents of a cone or pyramid.

---

* The area of a circle whose diameter is 1 = 0·7854.

RULE 2.

Multiply the area of the base by the perpendicular height, and one-third of the product will equal the solid contents.

To find the solid contents of a frustum of a cone.

RULE 3.

To the product of the diameters of the two ends add the sum of their squares; multiply this sum by the perpendicular height, and by ·2618;* the product equals contents.

To find the solid contents of a frustum of a pyramid.

RULE 4.

To the sum of the areas of the two ends add the square root of their product; multiply this sum by the perpendicular height, and one-third of the product equals the contents.

To find the solidity of a wedge.

RULE 5.

To the length of the wedge add twice the length of the base; multiply that sum by the height, and by the breadth of the base, and one-sixth of the product equals contents.

To find the solid contents of a prism.

RULE 6.

Multiply the area of the base by the length, equals contents.

---

* The solidity of a cone 1 foot diameter and 1 foot high equals ·2618.

To find the solid contents of a sphere or globe.

RULE 7.

Multiply the cube of the diameter by ·5236; the product equals contents.

To find the solid contents of the segment of a sphere.

RULE 8.

Add the square root of the height to three times the square of the radius of the base; multiply that sum by the height, and by ·5236; the product is the contents.

To find the solidity of a spheroid.

RULE 9.

Multiply the square of the least diameter by the length of the greatest diameter, or a line drawn perpendicular to the least diameter, and by ·5236; the product will be the solidity.

To find the solidity of a segment of a spheroid, when the base is circular or parallel to the revolving axis or least diameter.

RULE 10.

From triple the fixed axis take double the height of the segment; multiply the difference by the square of the height, and by ·5236. Then say, as the square of the fixed axis is to the square of the revolving axis, so is the former product to the solidity.

To find the solid contents of a cylindric ring.

RULE 11.

To the thickness of the ring add the inner diameter; multiply that sum by the square of the thickness, and by 2·4674; the product will be the solid contents.

To find the superficial contents of a board or plank.

RULE 12.

Multiply the length by the width.

If the plank or board are of an unequal breadth at the ends.

RULE 13.

Multiply the average width of the ends by the length.

To find the solidity of timber.

RULE 14.

Multiply the length in feet by the square of one-fourth the girth in inches, gives the solidity in cubic feet.

NOTE.—The above rules No. 12 and 13 only apply when all the dimensions are in feet. When either the length or breadth are given in inches, divide by 12, when all the dimensions are given in inches, divide by 144.

Application to the table of flat or board measure.

Multiply the length by the number in the table corresponding to any given width.

EXAMPLE.—Given a board $16\frac{1}{2}$ feet in length and $9\frac{3}{4}$ inches in breadth.

The number in the table opposite $9\frac{3}{4}$ inches $=$ $\cdot 8125 \times 16\frac{1}{2} = 13\cdot 4$ square feet.

TABLE TO FACILITATE THE MENSURATION OF TIMBER, FLAT OR BROAD MEASURE.

Breadth in inches.	Area of a Lineal foot.	Breadth in inches.	Area of a Lineal foot.	Breadth in inches.	Area of a Lineal foot.
¼	·0208	4½	·3750	8¾	·7292
½	·0417	4¾	·3958	9	·7500
¾	·0625	5	·4167	9¼	·7708
1	·0834	5¼	·4375	9½	·7917
1¼	·1042	5½	·4583	9¾	·8125
1½	·1250	4¾	·4792	10	·8334
1¾	·1459	6	·5000	10¼	·8542
2	·1667	6¼	·5208	10½	·8750
2¼	·1875	6½	·5416	10¾	·8959
2½	·2084	6¾	·5625	11	·9167
2¾	·2292	7	·5833	11¼	·9375
3	·2500	7¼	·6042	11½	·9583
3¼	·2708	7½	·6250	11¾	·9792
3½	·2916	7¾	·6458		
3¾	·3125	8	·6667		
4	·3334	8¼	·6875		
4¼	·3542	8½	·7084		

Application of the table of the solidity of timber. Multiply the area corresponding to the quarter girth in inches by the length in feet.

EXAMPLE.—Given a piece of timber 20 feet long and 12 inches square.

The number opposite 12 inches = 1·000 × 20 = 20 cubic feet.

TABLE TO FACILITATE THE MENSURATION OF THE SOLIDITY OF TIMBER.

1 qr girth in inches.	Area in feet.	1 qr. girth in inches.	Area in feet.	1 qr. girth in inches.	Area in feet.
6	·250	12¼	1·042	19	2·506
6¼	·272	12½	1·085	19½	2·640
6½	·294	12¾	1·129	20	2·777
6¾	·317	13	1·174	20½	2·917
7	·340	13¼	1·219	21	3·062
7¼	·364	13½	1 265	21½	3·209
7½	·390	13¾	1·313	22	3·362
7¾	·417	14	1·361	22½	3·516
8	·444	14¼	1·410	23	3 673
8¼	·472	14½	1·460	23½	3·835
8½	·501	14¾	1·511	24	4·000
8¾	·531	15	1·562	24½	4·168
9	·562	15¼	1·615	25	4·340
9¼	·594	15½	1·668	25½	4·516
9½	·626	15¾	1·722	26	4·694
9¾	·659	16	1·777	26½	4·876
10	·694	16¼	1·833	27	5·062
10¼	·730	16½	1·890	27½	5·252
10½	·766	16¾	1·948	28	5·444
10¾	·803	17	2·006	28½	5·640
11	·840	17¼	2·066	29	5·840
11¼	·878	17½	2·126	29½	6·044
11½	·918	17¾	2·187	30	6·250
11¾	·959	18	2·250		
12	1·000	18½	2·376		

Scantling is measured the same as timber, by multiplying the end area by the length.

MISCELLANIES.

The dimensions of the United States standard bushel are 18½ inches inside diameter, and 8 in. deep.

A box 24 inches by 16 inches square, and 28 inches deep, will contain a barrel, 5 bushels.

A box 14 inches by 17 inches square, and 14 inches deep, will contain a half barrel.

A box 26 inches by 15·2 inches square, and 8 inches deep, will contain 1 bushel.

A box 12 inches by 11·2 inches square, and 8 inches deep, will contain one-half bushel.

A box 8 inches by 8·4 inches square, and 8 inches deep, will contain 1 peck.

A box 8 inches by 8 inches square, and 4·2 inches deep, will contain 1 gallon.

A box 7 inches by 8 inches square, and 4·8 inches deep, will contain one-half gallon.

A box 4 inches by 4 inches square, and 4·1 inches deep, will contain 1 quart.

To get the number of bushels in any square crib.

RULE 1.

Find the number of cubic feet in the same, and multiply it by 8 and divide it by 10.

Any area in feet multiplied by 6·232, the product is the number of imperial gallons at one foot in depth; or any area in inches multiplied by 0·4328 = gallons.

Any area multiplied by ·03704 = the number of cubic yards at one foot in depth.

To determine the amount of imperial gallons in a vessel, the shape of an inverted cone.

RULE 2.

The square of the sum of the diameter at the top and bottom, of which subtract the quotient of the top and bottom; multiply the remainder by ·7854, and by one-third the depth = cubic feet; and by 6·232 = imperial gallons.

NOTE.— This rule applies where the dimensions are all given in feet.

2d. To the product of the inner diameters add the squares of the inner diameters; multiply the remainder by the depth, and by ·2618; divide that by 277·274 = gallons, (nearly.)

NOTE.— This rule applies where the dimensions are given in inches.

To determine the contents of imperial gallons in a kettle forming the segment of a circle.

RULE 3.

Three times the square of half the diameter in inches at the mouth, added to the square of the depth, and multiplied by the depth, and by ·5236; divide their product by 277·274, equals imperial gallons.

The area of a circle in inches, multiplied by the length or thickness in inches, and by ·263, equals the weight of cast iron in pounds.

The old English ale gallon contains 282 cubic inches, and the United States gallon contains 231.

English Dry Measure.

8·665	cubic inches	= 1 gill.
34·659	" "	= 1 pint.
69·318	" "	= 1 quart.
277·274	" "	= 1 gallon.
554·548	" "	= 1 peck.
1·2837	" feet	= 1 bushel.

English Imperial Wine Measure.

1·604	cubic feet	= 1 anker.
2·888	" "	= 1 runlet.
6·739	" "	= 1 tierce.
10·109	" "	= 1 hogshead.
3·478	" "	= 1 puncheon.
20·218	" "	= 1 pipe.
40·435	" "	= 1 tun.

Dimensions of Drawing Paper.

Wove Antiquarian,	4	feet	4	inches by	2	feet	7	inches.
Double Elephant,	3	"	4	"	2	"	2	"
Atlas,	2	"	9	"	2	"	2	"
Columbier,	2	"	9¾	"	1	"	11	"
Elephant,	2	"	3¾	"	1	"	10¼	"
Imperial,	2	"	5	"	1	"	9¼	"
Super Royal,	2	"	3	"	1	"	7	"
Royal,	2	"	0	"	1	"	7	"
Medium,	1	"	10	"	1	"	6	"
Demy,	1	"	7½	"	1	"	3½	"

Manner of Calculating the Natural Sines and Cosines in the Table.

The radius of a circle being 1, it is known the semi-circumference will equal 3·141592653589 8, &c. Therefore if we divide it by the number of minutes in a semi-circumference (10800) it will equal the sine of 1 minute = ·0002909 the first seven places in the table.

The natural cosine equals $\sqrt{(1-\text{sine}')}$ = ·999999-9577.

The natural sine and cosine given, the statement would be as follows:

2 cosine 1′ × sine 1′ — 0′ = sine 2′ minute.
2 " 1′ × " 2′ — 1′ = " 3′ "
2 " 1′ × " 3′ — 2′ = " 4′ "
2 " 1′ × " 4′ — 3′ = " 5′ "

This process can be run to any extent.

With the sine of one minute, and cosine of one minute given, statement second would be as follows:

1′ : sine 2′ — sine 1′ : : sine 2′ + sine 1′ : sine 3′
2′ : " 3′ — " 1′ : : " 3′ + " 1′ : " 4′
3′ : " 4′ — " 1′ : : " 4′ + " 1′ : " 5′
4′ : " 5′ — " 1′ : : " 5′ + " 1′ : " 6′

The same statement can be applied for degrees: Thus:

sine 1° : sine 2° — sine 1° : : sine 2° + sine 1° : sine 3°, &c., &c.

The natural sine of any number of degrees of deflexion with chords of 100 ft. may be found by dividing the chord by the radius corresponding to the angle of deflexion in the table of radii.

EXAMPLE.—Given the deflexion angle 3°, radius corresponding in the table of radii would equal 1910 feet and chord 100 feet.

STATEMENT.—100 ÷ 1910 = 0·052356, natural sine of 3°.

NATURAL SINES AND TANGENTS

TO A RADIUS 1.

NATURAL SINES AND TANGENTS TO A RADIUS 1.

0 Deg. 0 Deg.

′	SINE.	TANG.	COTANG.	COSINE.	′	′	SINE.	TANG.	COTANG.	COSINE.	′
0	·0000000	·000000	Infinite	1·000000	60	31	·0090174	·009017	110·8920	·9999593	29
1	·0002909	·000291	3437·746	1·000000	59	32	·0093083	·009308	107·4264	·9999567	28
2	·0005818	·000582	1718·873	·9999998	58	33	·0095992	·009599	104·1709	·9999539	27
3	·0008727	·000872	1145·915	·9999996	57	34	·0098900	·009890	101·1069	·9999511	26
4	·0011636	·001163	859·4363	·9999993	56	35	·0101809	·010181	98·21794	·9999482	25
5	·0014544	·001454	687·5488	·9999989	55	36	·0104718	·010472	95.48947	·9999452	24
6	·0017453	·001745	572·9572	·9999985	54	37	·0107627	·010763	92·90848	·9999421	23
7	·0020362	·002036	491·1060	·9999979	53	38	·0110535	·011054	90·46333	·9999389	22
8	·0023271	·002327	429·7175	·9999973	52	39	·0113444	·011345	88·14357	·9999357	21
9	·0026180	·002618	381·9709	·9999966	51	40	·0116353	·011636	85·93979	·9999323	20
10	·0029089	·002908	343·7737	·9999958	50	41	·0119261	·011927	83·84350	·9999289	19
11	·0031998	·003199	312·5213	·9999949	49	42	·0122170	·012217	81·84704	·9999254	18
12	·0034907	·003490	286·4777	·9999939	48	43	·0125079	·012508	79·94343	·9999218	17
13	·0037815	·003781	264·4408	·9999928	47	44	·0127987	·012799	78·12634	·9999181	16
14	·0040724	·004072	245·5519	·9999917	46	45	·0130896	·013090	76·39000	·9999143	15
15	·0043633	·004363	229·1816	·9999905	45	46	·0133805	·013381	74·72916	·9999105	14
16	·0046542	·004654	214·8576	·9999892	44	47	·0136713	·013672	73·13899	·9999065	13
17	.0049451	·004945	202·2187	.9999878	43	48	·0139622	·013963	71·61507	·9999025	12
18	·0052360	·005236	190·9841	·9999863	42	49	·0142530	·014254	70·15334	·9998984	11
19	·0055268	·005526	180·9322	·9999847	41	50	·0145439	·014545	68·75008	·9998942	10
20	·0058177	·005817	171·8854	·9999831	40	51	·0148348	·014836	67·40185	·9998900	9
21	·0061086	·006108	163·7001	·9999813	39	52	·0151256	·015127	66·10547	·9998856	8
22	·0063995	·006399	156·2590	·9999795	38	53	·0154165	·015418	64·85800	·9998812	7
23	·0066904	·006690	149·4650	·9999776	37	54	·0157073	·015709	63·65674	·9998766	6
24	·0069813	·006981	143·2371	·9999756	36	55	·0159982	·016000	62·49915	·9998720	5
25	·0072721	·007272	137·5075	·9999736	35	56	·0162890	·016291	61·38290	·9998673	4
26	·0075630	·007563	132·2185	·9999714	34	57	·0165799	·016582	60·30582	·9998625	3
27	·0078539	·007854	127·3213	·9999692	33	58	·0168707	·016873	59·26587	·9998577	2
28	·0081448	·008145	122·7739	·9999668	32	59	·0171616	·017164	58·26117	·9998527	1
29	·0084357	·008436	118·5401	·9999644	31	60	·0174524	·017455	57·28996	·9998477	0
30	·0087265	·008726	114·5886	·9999619	30						
′	COSINE.	COTANG.	TANG.	SINE.	′	′	COSINE.	COTANG.	TANG.	SINE.	′

Deg. 89. Deg. 89.

NATURAL SINES AND TANGENTS TO A RADIUS 1.

1 Deg.

′	SINE.	TANG.	COTANG.	COSINE.	′
0	·0174524	·017455	57·28996	·9998477	60
1	·0177432	·017746	56·35059	·9998426	59
2	·0180341	·018037	55·44151	·9998374	58
3	·0183249	·018328	54·56130	·9998321	57
4	·0186158	·018619	53·70858	·9998267	56
5	·0189066	·018910	52·88211	·9998213	55
6	·0191974	·019201	52·08067	·9998157	54
7	·0194883	.019492	51·30315	·9998101	53
8	·0197791	·019783	50·54850	·9998044	52
9	·0200699	·020074	49·81572	·9997986	51
10	·0203608	·020365	49·10388	·9997927	50
11	·0206516	·020656	48·41208	·9997867	49
12	·0209424	·020947	47·73950	·9997807	48
13	.0212332	·021238	47·08534	·9997745	47
14	·0215241	·021529	46·44886	·9997683	46
15	·0218149	·021820	45·82935	·9997620	45
16	·0221057	·022111	45·22614	·9997556	44
17	·0223965	·022402	44·63859	·9997492	43
18	·0226873	·022693	44·06611	·9997426	42
19	·0229781	·022984	43·50812	·9997360	41
20	·0232690	·023275	42·96407	·9997292	40
21	·0235598	·023566	42·43346	·9997224	39
22	·0238506	·023857	41·91579	·9997156	38
23	·0241414	·024148	41·41058	·9997086	37
24	·0244322	·024439	40·91741	·9997015	36
25	·0247230	·024730	40·43583	·9996943	35
26	·0250138	·025021	39·96546	·9996871	34
27	·0253046	·025312	39·50589	·9996798	33
28	·0255954	·025603	39·05677	·9996724	32
29	·0258862	·025894	38·61773	·9996649	31
30	·0261769	·026185	38·18845	·9996573	30
′	COSINE.	COTANG.	TANG.	SINE.	′

Deg 88.

1 Deg.

′	SINE.	TANG.	COTANG.	COSINE.	′
31	·0264677	·026477	37·76861	·9996497	29
32	·0267585	·026768	37·35789	·9996419	28
33	·0270493	·027059	36·95600	·9996341	27
34	·0273401	·027350	36·56265	·9996262	26
35	·0276309	·027641	36·17759	·9996182	25
36	·0279216	·027932	35·80055	·9996101	24
37	·0282124	·028223	35·43123	·9996020	23
38	·0285032	·028514	35·06954	·9995937	22
39	·0287940	·028805	34·71511	·9995854	21
40	·0290847	·029097	34·36777	·9995770	20
41	·0293755	·029388	34·02730	·9995684	19
42	·0296662	·029679	33·69350	·9995599	18
43	·0299570	·029970	33·36619	·9995512	17
44	·0302478	·030261	33·04517	·9995424	16
45	·0305385	·030552	32·73026	·9995336	15
46	·0308293	·030843	32·42129	·9995247	14
47	·0311200	·031135	32·11809	·9995157	13
48	.0314108	·031426	31·82051	·9995066	12
49	·0317015	·031717	31·52839	·9994974	11
50	·0319922	·032008	31·24157	·9994881	10
51	·0322830	·032299	30·95992	·9994788	9
52	·0325737	·032591	30·68330	·9994693	8
53	·0328644	·032882	30·41158	·9994598	7
54	·0331552	·033173	30·14461	·9994502	6
55	·0334459	·033464	29·88229	·9994405	5
56	·0337366	·033755	29·62449	·9994308	4
57	·0340274	.034047	29·37110	·9994209	3
58	·0343181	·034338	29·12200	·9994110	2
59	·0346088	·034629	28·87708	·9994009	1
60	·0348995	·034920	28·63625	·9993908	0
′	COSINE.	COTANG.	TANG.	SINE.	′

Deg. 88.

NATURAL SINES AND TANGENTS TO A RADIUS 1.

2 Deg. | 2 Deg.

′	SINE.	TANG.	COTANG.	COSINE.	′	′	SINE.	TANG.	COTANG.	COSINE.	′
0	·0348995	·034920	28·63625	·9993908	60	31	·0439100	·043952	22·75189	·9990355	29
1	·0351902	·035212	28·39939	·9993806	59	32	·0442006	·044243	22·60201	·9990227	28
2	·0354809	·035503	28·16642	·9993704	58	33	·0444912	·044535	22·45409	·9990098	27
3	·0357716	·035794	27·93723	·9993600	57	34	·0447818	·044826	22·30809	·9989968	26
4	·0360623	·036085	27·71174	·9993495	56	35	·0450724	·045118	22·16398	·9989837	25
5	·0363530	·036377	27·48985	·9993390	55	36	·0453630	·045409	22·02171	·9989706	24
6	·0366437	·036668	27·27148	·9993284	54	37	·0456536	·045701	21·88125	·9989573	23
7	·0369344	·036959	27·05655	·9993177	53	38	·0459442	·045992	21·74256	·9989440	22
8	·0372251	·037250	26·84498	·9993069	52	39	·0462347	·046284	21·60563	·9989306	21
9	·0375158	·037542	26·63669	·9992960	51	40	·0465253	·046575	21·47040	·9989171	20
10	·0378065	·037833	26·43160	·9992851	50	41	·0468159	·046867	21·33685	·9989035	19
11	·0380971	·038124	26·22963	·9992740	49	42	·0471065	·047158	21·20494	·9988899	18
12	·0383878	·038416	26·03073	·9992629	48	43	·0473970	·047450	21·07466	·9988761	17
13	·0386785	·038707	25·83482	·9992517	47	44	·0476876	·047741	20·94596	·9988623	16
14	·0389692	·038998	25·64183	·9992404	46	45	·0479781	·048033	20·81882	·9988484	15
15	·0392598	·039290	25·45170	·9992290	45	46	·0482687	·048325	20·69322	·9988344	14
16	·0395505	·039581	25·26436	·9992176	44	47	·0485592	·048616	20·56911	·9988203	13
17	·0398411	·039872	25·07975	·9992060	43	48	·0488498	·048908	20·44648	·9988061	12
18	·0401318	·040164	24·89782	·9991944	42	49	·0491403	·049199	20·32530	·9987919	11
19	·0404224	·040455	24·71851	·9991827	41	50	·0494308	·049491	20·20555	·9987775	10
20	·0407131	·040746	24·54175	·9991709	40	51	·0497214	·049782	20·08719	·9987631	9
21	·0410037	·041038	24·36750	·9991590	39	52	·0500119	·050074	19·97021	·9987486	8
22	·0412944	·041329	24·19571	·9991470	38	53	·0503024	·050366	19·85459	·9987340	7
23	·0415850	·041621	24·02632	·9991350	37	54	·0505929	·050657	19·74029	·9987194	6
24	·0418757	·041912	23·85927	·9991228	36	55	·0508835	·050949	19·62729	·9987046	5
25	·0421663	·042203	23·69453	·9991106	35	56	·0511740	·051241	19·51558	·9986898	4
26	·0424569	·042495	23·53205	·9990983	34	57	·0514645	·051532	19·40513	·9986748	3
27	·0427475	·042786	23·37177	·9990859	33	58	·0517550	·051824	19·29592	·9986598	2
28	·0430382	·043078	23·21366	·9990734	32	59	·0520455	·052116	19·18793	·9986447	1
29	·0433288	·043369	23·05767	·9990609	31	60	·0523360	·052407	19·08113	·9986295	0
30	·0436194	·043660	22·90376	·9990482	30						
′	COSINE.	COTANG.	TANG.	SINE.	′	′	COSINE.	COTANG.	TANG.	SINE.	′

Deg. 87. | Deg. 87.

NATURAL SINES AND TANGENTS TO A RADIUS 1.

3 Deg.

′	SINE.	TANG.	COTANG.	COSINE.	′
0	·0523360	·052407	19·08113	·9986295	60
1	·0526264	·052699	18·97552	·9986143	59
2	·0529169	·052991	18·87106	·9985989	58
3	·0532074	·053282	18·76775	·9985835	57
4	·0534979	·053574	18·66556	·9985680	56
5	·0537883	·053866	18·56447	·9985524	55
6	·0540788	·054158	18·46447	·9985367	54
7	·0543693	·054449	18·36553	·9985209	53
8	·0546597	·054741	18·26765	·9985050	52
9	·0549502	·055033	18·17080	·9984891	51
10	·0552406	·055325	18·07497	·9984731	50
11	·0555311	·055616	17·98015	·9984570	49
12	·0558215	·055908	17·88631	·9984408	48
13	·0561119	·056200	17·79344	·9984245	47
14	·0564024	·056492	17·70152	·9984081	46
15	·0566928	·056784	17·61055	·9983917	45
16	·0569832	·057075	17·52051	·9983751	44
17	·0572736	·057367	17·43138	·9983585	43
18	·0575640	·057659	17·34315	·9983418	42
19	·0578544	·057951	17·25580	·9983250	41
20	·0581448	·058243	17·16933	·9983082	40
21	·0584352	·058535	17·08372	·9982912	39
22	·0587256	·058827	16·99895	·9982742	38
23	·0590160	·059119	16·91502	·9982570	37
24	·0593064	·059410	16·83191	·9982398	36
25	·0595967	·059702	16·74961	·9982225	35
26	·0598871	·059994	16·66811	·9982052	34
27	·0601775	·060286	16·58739	·9981877	33
28	·0604678	·060578	16·50745	·9981701	32
29	·0607582	·060870	16·42827	·9981525	31
30	·0610485	·061162	16·34985	·9981348	30
′	COSINE.	COTANG.	TANG.	SINE.	′

Deg. 86.

3 Deg.

′	SINE.	TANG.	COTANG.	COSINE.	′
31	·0613389	·064154	16·27217	·9981170	29
32	·0616292	·061746	16·19522	·9980991	28
33	·0619196	·062038	16·11899	·9980811	27
34	·0622099	·062330	16·04348	·9980631	26
35	·0625002	·062622	15·96866	·9980450	25
36	·0627905	·062914	15·89454	·9980267	24
37	·0630808	·063206	15·82110	·9980084	23
38	·0633711	·063498	15·74833	·9979900	22
39	·0636614	·063790	15·67623	·9979716	21
40	·0639517	·064082	15·60478	·9979530	20
41	·0642420	·064375	15·53398	·9979343	19
42	·0645323	·064667	15·46381	·9979156	18
43	·0648226	·064959	15·39427	·9978968	17
44	·0651129	·065251	15·32535	·9978779	16
45	·0654031	·065543	15·25705	·9978589	15
46	·0656934	·065835	15·18934	·9978399	14
47	·0659836	·066127	15·12224	·9978207	13
48	·0662739	·066419	15·05572	·9978015	12
49	·0665641	·066712	14·98978	·9977821	11
50	·0668544	·067004	14·92441	·9977627	10
51	·0671446	·067296	14·85961	·9977433	9
52	·0674349	·067588	14·79537	·9977237	8
53	·0677251	·067880	14·73167	·9977040	7
54	·0680153	·068173	14·66852	·9976843	6
55	·0683055	·068465	14·60591	·9976645	5
56	·0685957	·068757	14·54383	·9976445	4
57	·0688859	·069049	14·48227	·9976245	3
58	·0691761	·069342	14·42123	·9976045	2
59	·0694663	·069634	14·36069	·9975843	1
60	·0697565	·069926	14·30066	·9975641	0
′	COSINE.	COTANG.	TANG.	SINE.	′

Deg. 86.

NATURAL SINES AND TANGENTS TO A RADIUS 1.

4 Deg. | 4 Deg.

′	SINE.	TANG.	COTANG.	COSINE.	′	′	SINE.	TANG.	COTANG.	COSINE.	′
0	·0697565	·069926	14·30066	·9975641	60	31	·0787491	·078994	12·65912	·9968945	29
1	·0700467	·070219	14·24113	·9975437	59	32	·0790391	·079287	12·61239	·9968715	28
2	·0703368	·070511	14·18209	·9975233	58	33	·0793290	·079579	12·56599	·9968485	27
3	·0706270	·070803	14·12353	·9975028	57	34	·0796190	·079872	12·51994	·9968254	26
4	·0709171	·071096	14·06545	·9974822	56	35	·0799090	·080165	12·47422	·9968022	25
5	·0712073	·071388	14·00785	·9974615	55	36	·0801989	·080458	12·42883	·9967789	24
6	·0714974	·071680	13·95071	·9974408	54	37	·0804889	·080750	12·38376	·9967555	23
7	·0717876	·071973	13·89404	·9974199	53	38	·0807788	·081043	12·33902	·9967321	22
8	·0720777	·072265	13·83782	·9973990	52	39	·0810687	·081336	12·29460	·9967085	21
9	·0723678	·072558	13·78206	·9973780	51	40	·0813587	·081629	12·25050	·9966849	20
10	·0726580	·072850	13·72673	·9973569	50	41	·0816486	·081922	12·20671	·9966612	19
11	·0729481	·073143	13·67185	·9973357	49	42	·0819385	·082215	12·16323	·9966374	18
12	·0732382	·073425	13·61740	·9973145	48	43	·0822284	·082507	12·12006	·9966135	17
13	·0735283	·073727	13·56339	·9972931	47	44	0825183	·082800	12·07719	·9965895	16
14	·0738184	·074020	13·50979	·9972717	46	45	·0828082	·083093	12·03462	·9965655	15
15	·0741085	·074312	13·45662	·9972502	45	46	·0830981	·083386	11·99234	·9965414	14
16	·0743986	·074605	13·40386	·9972286	44	47	·0833880	·083679	11·95037	·9965172	13
17	·0746887	·074897	13·35151	·9972069	43	48	·0836778	·083972	11·90868	·9964929	12
18	·0749787	·075190	13·29957	·9971851	42	49	·0839677	·084265	11·86728	·9964685	11
19	·0752688	·075482	13·24803	·9971633	41	50	·0842576	·084558	11·82616	·9964440	10
20	·0755589	·075775	13·19688	·9971413	40	51	·0845474	·084851	11·78533	·9964195	9
21	·0758489	·076068	13·14612	·9971193	39	52	·0848373	·085144	11·74477	·9963948	8
22	·0761390	·076360	13.09575	·9970972	38	53	·0851271	·085437	11·70450	·9963701	7
23	·0764290	·076653	13·04576	·9970750	37	54	·0854169	·085730	11·66449	·9963453	6
24	·0767190	·076945	12·99616	·9970528	36	55	·0857067	·086023	11·62476	·9963204	5
25	·0770091	·077238	12·94692	·9970304	35	56	·0859966	·086316	11·58529	·9962954	4
26	·0772991	·077531	12·89805	·9970080	34	57	·0862864	·086609	11·54609	·9962704	3
27	·0775891	·077823	12·84955	·9969854	33	58	·0865762	·086902	11·50715	·9962452	2
28	·0778791	·078116	12·80141	·9969628	32	59	·0868660	·087195	11·46847	·9962200	1
29	·0781691	·078409	12·75363	·9969401	31	60	·0871557	·087488	11·43005	·9961947	0
30	·0784591	·078701	12·70620	·9969173	30						
′	COSINE.	COTANG.	TANG.	SINE.	′	′	COSINE.	COTANG.	TANG.	SINE.	′

Deg. 85. | Deg. 85.

NATURAL SINES AND TANGENTS TO A RADIUS 1.

5 Deg.

′	SINE.	TANG.	COTANG.	COSINE.	′
0	·0871557	·087488	11·43005	·9961947	60
1	·0874455	·087781	11·39188	·9961693	59
2	·0877353	·088074	11·35397	·9961438	58
3	·0880251	·088368	11·31630	·9961183	57
4	·0883148	·088661	11·27888	·9960926	56
5	·0886046	·088954	11·24171	·9960669	55
6	·0888943	·089247	11·20478	·9960411	54
7	·0891840	·089540	11·16808	·9960152	53
8	·0894738	·089834	11·13163	·9959892	52
9	·0897635	·090127	11·09541	·9959631	51
10	·0900532	·090420	11·05943	·9959370	50
11	·0903429	·090713	11·02367	·9959107	49
12	·0906326	·091007	10·98815	·9958844	48
13	·0909223	·091300	10·95285	·9958580	47
14	·0912119	·091593	10·91777	·9958315	46
15	·0915016	·091887	10·88292	·9958049	45
16	·0917913	·092180	10·84828	·9957783	44
17	·0920809	·092473	10·81387	·9957515	43
18	·0923706	·092767	10·77967	·9957247	42
19	·0926602	·093060	10·74568	·9956978	41
20	·0929499	·093354	10·71191	·9956708	40
21	·0932395	·093647	10·67834	·9956437	39
22	·0935291	·093940	10·64499	·9956165	38
23	·0938187	·094234	10·61184	·9955892	37
24	·0941083	·094527	10·57889	·9955620	36
25	·0943979	·094821	10·54615	·9955345	35
26	·0946875	·095114	10·51360	·9955070	34
27	·0949771	·095408	10·48126	·9954794	33
28	·0952666	·095701	10·44911	·9954517	32
29	·0955562	·095995	10·41715	·9954240	31
30	·0958458	·096289	10·38539	·9953962	30
′	COSINE.	COTANG.	TANG.	SINE.	′

Deg. 84.

5 Deg.

′	SINE.	TANG.	COTANG.	COSINE.	′
31	·0961353	·096582	10·35382	·9953683	29
32	·0964248	·096876	10·32244	·9953403	28
33	·0967144	·097169	10·29125	·9953122	27
34	·0970039	·097463	10·26024	·9952840	26
35	·0972934	·097757	10·22942	·9952557	25
36	·0975829	·098050	10·19878	·9952274	24
37	·0978724	·098344	10·16833	·9951990	23
38	·0981619	·098638	10·13805	·9951705	22
39	·0983514	·098932	10·10795	·9951419	21
40	·0987408	·099225	10·07803	·9951132	20
41	·0990303	·099519	10·04828	·9950844	19
42	·0993197	·099813	10·01871	·9950556	18
43	·0996092	·100107	9·989305	·9950266	17
44	·0998986	·100400	9·960072	·9949976	16
45	·1001881	·100694	9·931008	·9949685	15
46	·1004775	·100988	9·902112	·9949393	14
47	·1007669	·101282	9·873382	·9949101	13
48	·1010563	·101576	9·844816	·9948807	12
49	·1013457	·101870	9·816414	·9948513	11
50	·1016351	·102164	9·788173	·9948217	10
51	·1019245	·102458	9·760092	·9947921	9
52	·1022138	·102752	9·732171	·9947625	8
53	·1025032	·103046	9·704407	·9947327	7
54	·1027925	·103339	9·676800	·9947028	6
55	·1030819	·103634	9·649347	·9946729	5
56	·1033712	·103928	9·622048	·9946428	4
57	·1036605	·104222	9·594902	·9946127	3
58	·1039499	·104516	9·567906	·9945825	2
59	·1042392	·104810	9·541061	·9945523	1
60	·1045285	·105104	9·514364	·9945219	0
′	COSINE.	COTANG.	TANG.	SINE.	′

Deg. 84.

NATURAL SINES AND TANGENTS TO A RADIUS 1.

6 Deg. | 6 Deg.

′	SINE.	TANG.	COTANG.	COSINE.	′	′	SINE.	TANG.	COTANG.	COSINE.	′
0	·1045285	·105104	9·514364	·9945219	60	31	·1134922	·114230	8·754246	·9935389	29
1	·1048178	·105398	9·487814	·9944914	59	32	·1137812	·114525	8·731719	·9935058	28
2	·1051070	·105692	9·461411	·9944609	58	33	·1140702	·114819	8·709307	·9934727	27
3	·1053963	·105986	9·435153	·9944303	57	34	·1143592	·115114	8·687008	·9934395	26
4	·1056856	·106280	9·409038	·9943996	56	35	·1146482	·115409	8·664822	·9934062	25
5	·1059748	·106575	9·383066	·9943688	55	36	·1149372	·115703	8·642747	·9933728	24
6	·1062641	·106869	9·357235	·9943379	54	37	·1152261	·115998	8·620783	·9933393	23
7	·1065533	·107163	9·331545	·9943070	53	38	·1155151	·116293	8·598929	·9933057	22
8	·1068425	·107457	9·305993	·9942760	52	39	·1158040	·116588	8·577183	·9932721	21
9	·1071318	·107751	9·280580	·9942448	51	40	·1160929	·116883	8·555546	·9932384	20
10	·1074210	·108046	9·255303	·9942136	50	41	·1163818	·117178	8·534017	·9932045	19
11	·1077102	·108340	9·230162	·9941823	49	42	·1166707	·117473	8·512594	·9931706	18
12	·1079994	·108634	9·205156	·9941510	48	43	·1169596	·117767	8·491277	·9931367	17
13	·1082885	·108929	9·180283	·9941195	47	44	·1172485	·118062	8·470065	·9931026	16
14	·1085777	·109223	9·155543	·9940880	46	45	·1175374	·118357	8·448957	·9930685	15
15	·1088669	·109517	9·130934	·9940563	45	46	·1178263	·118652	8·427953	·9930342	14
16	·1091560	·109812	9·106456	·9940246	44	47	·1181151	·118947	8·407051	·9929999	13
17	·1094452	·110106	9·082107	·9939928	43	48	·1184040	·119242	8·386251	·9929655	12
18	·1097343	·110401	9·057886	·9939610	42	49	·1186928	·119537	8·365553	·9929310	11
19	·1100234	·110695	9·033793	·9939290	41	50	·1189816	·119832	8·344955	·9928965	10
20	·1103126	·110989	9·009826	·9938969	40	51	·1192704	·120127	8·324457	·9928618	9
21	·1106017	·111284	8·985984	·9938648	39	52	·1195593	·120423	8·304058	·9928271	8
22	·1108908	·111578	8·962266	·9938326	38	53	·1198481	·120718	8·283757	·9927922	7
23	·1111779	·111873	8·938672	·9938003	37	54	·1201368	·121013	8·263554	·9927573	6
24	·1114689	·112168	8·915200	·9937679	36	55	·1204256	·121308	8·243448	·9927224	5
25	·1117580	·112462	8·891850	·9937355	35	56	·1207144	·121603	8·223438	·9926873	4
26	·1120471	·112757	8·868620	·9937029	34	57	·1210031	·121898	8·203523	·9926521	3
27	·1123361	·113051	8·845510	·9936703	33	58	·1212919	·122194	8·183704	·9926169	2
28	·1126252	·113346	8·822518	·9936375	32	59	·1215806	·122489	8·163978	·9925816	1
29	·1129142	·113641	8·799644	·9936047	31	60	·1218693	·122784	8·144346	·9925462	0
30	·1132032	·113935	8·776887	·9935719	30						
′	COSINE.	COTANG.	TANG.	SINE.	′	′	COSINE.	COTANG.	TANG.	SINE.	′

Deg. 83. | Deg. 83

NATURAL SINES AND TANGENTS TO A RADIUS 1.

7 Deg. 7 Deg.

′	SINE.	TANG.	COTANG.	COSINE.	′	′	SINE.	TANG.	COTANG.	COSINE.	′
0	·1218693	·122784	8·144346	·9925462	60	31	·1308146	·131948	7·578717	·9914069	29
1	·1221581	·123079	8·124807	·9925107	59	32	·1311030	·132244	7·561756	·9913688	28
2	·1224468	·123375	8·105359	·9924751	58	33	·1313913	·132540	7·544869	·9913306	27
3	·1227355	·123670	8·086004	·9924394	57	34	·1316797	·132836	7·528057	·9912923	26
4	·1230241	·123965	8·066739	·9924037	56	35	·1319681	·133132	7·511317	·9912540	25
5	·1233128	·124261	8·047564	·9923679	55	36	·1322564	·133428	7·494651	·9912155	24
6	·1236015	·124556	8·028479	·9923319	54	37	·1325447	·133724	7·478057	·9911770	23
7	·1238901	·124852	8·009483	·9922959	53	38	·1328330	·134020	7·461535	·9911384	22
8	·1241788	·125147	7·990575	·9922599	52	39	·1331213	·134316	7·445085	·9910997	21
9	·1244674	·125442	7·971755	·9922237	51	40	·1334096	·134612	7·428706	·9910610	20
10	·1247560	·125738	7·953022	·9921874	50	41	·1336979	·134909	7·412397	·9910221	19
11	·1250446	·126033	7·934375	·9921511	49	42	·1339862	·135205	7·396159	·9909832	18
12	·1253332	·126329	7·915815	·9921147	48	43	·1342744	·135501	7·379990	·9909442	17
13	·1256218	·126624	7·897339	·9920782	47	44	·1345627	·135797	7·363891	·9909051	16
14	·1259104	·126920	7·878948	·9920416	46	45	·1348509	·136094	7·347861	·9908659	15
15	·1261990	·127216	7·860642	·9920049	45	46	·1351392	·136390	7·331898	·9908266	14
16	·1264875	·127511	7·842419	·9919682	44	47	·1354274	·136686	7·316004	·9907873	13
17	·1267761	·127807	7·824279	·9919314	43	48	·1357156	·136983	7·300178	·9907478	12
18	·1270646	·128103	7·806221	·9918944	42	49	·1360038	·137279	7·284418	·9907083	11
19	·1273531	·128398	7·788245	·9918574	41	50	·1362919	·137575	7·268725	·9906687	10
20	·1276416	·128694	7·770350	·9918204	40	51	·1365801	·137872	7·253098	·9906290	9
21	·1279302	·128990	7·752536	·9917832	39	52	·1368683	·138168	7·237537	·9905893	8
22	·1282186	·129285	7·734802	·9917459	38	53	·1371564	·138465	7·222042	·9905494	7
23	·1285071	·129581	7·717148	·9917086	37	54	·1374445	·138761	7·206611	·9905095	6
24	·1287956	·129877	7·699573	·9916712	36	55	·1377327	·139058	7·191245	·9904694	5
25	·1290841	·130173	7·682076	·9916337	35	56	·1380208	·139354	7·175943	·9904293	4
26	·1293725	·130469	7·664658	·9915961	34	57	·1383089	·139651	7·160705	·9903891	3
27	·1296609	·130764	7·647317	·9915584	33	58	·1385970	·139947	7·145530	·9903489	2
28	·1299494	·131060	7·630053	·9915206	32	59	·1388850	·140244	7·130419	·9903085	1
29	·1302378	·131356	7·612865	·9914828	31	60	·1391731	·140540	7·115369	·9902681	0
30	·1305262	·131652	7·595754	·9914449	30						
′	COSINE.	COTANG.	TANG.	SINE.	′	′	COSINE.	COTANG.	TANG.	SINE.	′

Deg. 82. Deg. 82.

NATURAL SINES AND TANGENTS TO A RADIUS 1.

8 Deg.						8 Deg.					
′	SINE.	TANG.	COTANG.	COSINE.	′	′	SINE.	TANG.	COTANG.	COSINE.	′
0	·1391731	·140540	7·115369	·9902681	60	31	·1480971	·149748	6·677867	·9889728	29
1	·1394612	·140837	7·100382	·9902275	59	32	·1483848	·150045	6·664630	·9889297	28
2	·1397492	·141134	7·085457	·9901869	58	33	·1486724	·150343	6·651444	·9888895	27
3	·1400372	·141430	7·070593	·9901462	57	34	·1489601	·150640	6·638310	·9888432	26
4	·1403252	·141727	7·055790	·9901055	56	35	·1492477	·150938	6·625225	·9887998	25
5	·1406132	·142024	7·041048	·9900646	55	36	·1495353	·151235	6·612191	·9887564	24
6	·1409012	·142321	7·026366	·9900237	54	37	·1498230	·151533	6·599208	·9887128	23
7	·1411892	·142617	7·011744	·9899826	53	38	·1501106	·151830	6·586273	·9886692	22
8	·1414772	·142914	6·997180	·9899415	52	39	·1503981	·152128	6·573389	·9886255	21
9	·1417651	·143211	6·982678	·9899003	51	40	·1506857	·152426	6·560553	·9885817	20
10	·1420531	·143508	6·968233	·9898590	50	41	·1509733	·152723	6·547767	·9885378	19
11	·1423410	·143805	6·953847	·9898177	49	42	·1512608	·153021	6·535029	·9884939	18
12	·1426289	·144102	6·939519	·9897762	48	43	·1515484	·153319	6·522339	·9884498	17
13	·1429168	·144399	6·925248	·9897347	47	44	·1518359	·153617	6·509698	·9884057	16
14	·1432047	·144696	6·911035	·9896931	46	45	·1521234	·153914	6·497104	·9883615	15
15	·1434926	·144993	6·896879	·9896514	45	46	·1524109	·154212	6·484558	·9883172	14
16	·1437805	·145290	6·882780	·9896096	44	47	·1526984	·154510	6·472059	·9882728	13
17	·1440684	·145587	6·868737	·9895677	43	48	·1529858	·154808	6·459607	·9882284	12
18	·1443562	·145884	6·854750	·9895258	42	49	·1532733	·155106	6·447201	·9881838	11
19	·1446440	·146181	6·840819	·9894838	41	50	·1535607	·155404	6·434842	·9881392	10
20	·1449319	·146478	6·826943	·9894416	40	51	·1538482	·155701	6·422530	·9880945	9
21	·1452197	·146775	6·813122	·9893994	39	52	·1541356	·155999	6·410263	·9880497	8
22	·1455075	·147072	6·799356	·9893572	38	53	·1544230	·156297	6·398042	·9880048	7
23	·1457953	·147369	6·785644	·9893148	37	54	·1547104	·156595	6·385866	·9879599	6
24	·1460830	·147667	6·791986	·9892723	36	55	·1549978	·156893	6·373735	·9879148	5
25	·1463708	·147964	6·758382	·9892298	35	56	·1552851	·157191	6·361650	·9878697	4
26	·1466585	·148261	6·744831	·9891872	34	57	·1555725	·157490	6·349609	·9878245	3
27	·1469463	·148559	6·731334	·9891445	33	58	·1558598	·157788	6·337612	·9877792	2
28	·1472340	·148856	6·717889	·9891017	32	59	·1561472	·158086	6·325660	·9877338	1
29	·1475217	·149153	6·704496	·9890588	31	60	·1564345	·158384	6·313751	·9876883	0
30	·1478094	·149451	6·691156	·9890159	30						
′	COSINE.	COTANG.	TANG.	SINE.	′	′	COSINE.	COTANG.	TANG.	SINE.	′
				Deg. 81.						Deg. 81.	

NATURAL SINES AND TANGENTS TO A RADIUS 1.

9 Deg. | 9 Deg.

′	SINE.	TANG.	COTANG.	COSINE.	′	′	SINE.	TANG.	COTANG.	COSINE.	′
0	·1564345	·158384	6·313751	·9876883	60	31	·1653345	·167641	5·965104	·9862375	29
1	·1567218	·158682	6·301886	·9876428	59	32	·1656214	·167940	5·954481	·9861894	28
2	·1570091	·158980	6·290065	·9875972	58	33	·1659082	·168239	5·943895	·9861412	27
3	·1572963	·159279	6·278286	·9875514	57	34	·1661951	·168539	5·933345	·9860929	26
4	·1575836	·159577	6·266551	·9875057	56	35	·1664819	·168838	5·922832	·9860445	25
5	·1578708	·159875	6·254858	·9874598	55	36	·1667687	·169137	5·912355	·9859960	24
6	·1581581	·160174	6·243208	·9874138	54	37	·1670556	·169436	5·901913	·9859475	23
7	·1584453	·160472	6·231600	·9873678	53	38	·1673423	·169735	5·891508	·9858988	22
8	·1587325	·160770	6·220034	·9873216	52	39	·1676291	·170035	5·881138	·9858501	21
9	·1590197	·161069	6·208510	·9872754	51	40	·1679159	·170334	5·870804	·9858013	20
10	·1593069	·161367	6·197027	·9872291	50	41	·1682026	·170633	5·860505	·9857524	19
11	·1595940	·161666	6·185586	·9871827	49	42	·1684894	·170933	5·850241	·9857035	18
12	·1598812	·161964	6·174186	·9871363	48	43	·1687761	·171232	5·840011	·9856544	17
13	·1601683	·162263	6·162827	·9870897	47	44	·1690628	·171532	5·829817	·9856053	16
14	·1604555	·162561	6·151508	·9870431	46	45	·1693495	·171831	5·819657	·9855561	15
15	·1607426	162860	6·140230	·9869964	45	46	·1696362	·172130	5·809531	·9855068	14
16	·1610297	·163159	6·128992	·9869496	44	47	·1699228	·172430	5·799440	·9854574	13
17	·1613167	·163457	6·117794	·9869027	43	48	·1702095	·172730	5·789382	·9854079	12
18	·1616038	·163756	6·106636	·9868557	42	49	·1704961	·173029	5·779358	·9853583	11
19	·1618909	·164055	6·095517	·9868087	41	50	·1707828	·173329	5·769368	·9853087	10
20	·1621779	·164353	6·084438	·9867615	40	51	·1710694	·173628	5·759412	·9852590	9
21	·1624650	·164652	6·073397	·9867143	39	52	·1713560	·173928	5·749488	·9852092	8
22	·1627520	·164951	6·062396	·9866670	38	53	·1716425	·174228	5·739598	·9851593	7
23	·1630390	·165250	6·051434	·9866196	37	54	·1719291	·174527	5·729741	·9851093	6
24	·1633260	·165548	6·040510	·9865722	36	55	·1722156	·174827	5·719917	·9850593	5
25	·1636129	·165847	6·029624	·9865246	35	56	·1725022	·175127	5·710125	·9850091	4
26	·1638999	·166146	6·018777	·9864770	34	57	·1727887	·175427	5·700366	·9849589	3
27	·1641868	·166445	6·007967	·9864293	33	58	·1730752	·175727	5·690639	·9849086	2
28	·1644738	·166744	5·997195	·9863815	32	59	·1733617	·176027	5·680944	·9848582	1
29	·1647607	·167043	5·986461	·9863336	31	60	·1736482	·176327	5·671281	·9848078	0
30	·1650476	·167342	5·975764	·9862856	30						
′	COSINE.	COTANG.	TANG.	SINE.	′	′	COSINE.	COTANG.	TANG.	SINE.	′

Deg. 80. | Deg. 80.

NATURAL SINES AND TANGENTS TO A RADIUS 1.

10 Deg.

′	SINE.	TANG.	COTANG.	COSINE.	′
0	·1736482	·176327	5·671281	·9848078	60
1	·1739346	·176626	5·661650	·9847572	59
2	·1742211	·176926	5·652051	·9847066	58
3	·1745075	·177226	5·642483	·9846558	57
4	·1747939	·177527	5·632947	·9846050	56
5	·1750803	·177827	5·623442	·9845542	55
6	·1753667	·178127	5·613968	·9845032	54
7	·1756531	·178427	5·604524	·9844521	53
8	·1759395	·178727	5·595112	·9844010	52
9	·1762258	·179027	5·585730	·9843498	51
10	·1765121	·179327	5.576378	·9842985	50
11	·1767984	·179628	5·567057	·9842471	49
12	·1770847	·179928	5·557766	·9841956	48
13	·1773710	·180228	5·548505	·9841441	47
14	·1776573	·180529	5·539274	·9840924	46
15	·1779435	·180829	5·530072	·9840407	45
16	·1782298	·181129	5·520900	·9839889	44
17	·1785160	·181430	5·511757	·9839370	43
18	·1788022	·181730	5·502644	·9838850	42
19	·1790884	·182031	5·493560	·9838330	41
20	·1793746	·182331	5·484505	·9837808	40
21	·1796607	·182632	5·475478	·9837286	39
22	·1799469	·182933	5·466481	·9836763	38
23	·1802330	·183233	5·457512	·9836239	37
24	·1805191	·183534	5·448571	·9835715	36
25	·1808052	·183835	5·439659	·9835189	35
26	·1810913	·184135	5·430775	·9834663	34
27	·1813774	·184436	5·421918	·9834136	33
28	·1816635	·184737	5·413090	·9833608	32
29	·1819495	·185038	5·404290	·9833079	31
30	·1822355	·185339	5·395517	·9832549	30
′	COSINE.	COTANG.	TANG.	SINE.	′

Deg. 79.

10 Deg.

′	SINE.	TANG.	COTANG.	COSINE.	′
31	·1825215	·185639	5·386771	·9832019	29
32	·1828075	·185940	5·378053	·9831487	28
33	·1830935	·186241	5·369363	·9830955	27
34	·1833795	·186542	5·360699	·9830422	26
35	·1836654	·186843	5·352062	·9829888	25
36	·1839514	·187144	5·343452	·9829353	24
37	·1842373	·187446	5·334869	·9828818	23
38	·1845232	·187747	5·326313	·9828282	22
39	·1848091	·188048	5·317783	·9827744	21
40	·1850949	·188349	5·309279	·9827206	20
41	·1853808	·188650	5·300801	·9826668	19
42	·1856666	·188952	5·292350	·9826128	18
43	·1859524	·189253	5·283925	·9825587	17
44	·1862382	·189554	5·275525	·9825046	16
45	·1865240	·189855	5·267151	·9824504	15
46	·1868098	·190157	5·258803	·9823961	14
47	·1870956	·190458	5·250480	·9823417	13
48	·1873813	·190760	5·242183	·9822873	12
49	·1876670	·191061	5·233911	·9822327	11
50	·1879528	·191363	5·225664	·9821781	10
51	·1882385	·191664	5·217442	·9821234	9
52	·1885241	·191966	5·209245	·9820686	8
53	·1888098	·192268	5·201073	·9820137	7
54	·1890954	·192569	5·192926	·9819587	6
55	·1893811	·192871	5·184803	·9819037	5
56	·1896667	·193173	5·176705	·9818485	4
57	·1899523	·193474	5·168631	·9817933	3
58	·1902379	·193776	5·160581	·9817380	2
59	·1905234	·194078	5·152555	·9816826	1
60	·1908090	·194380	5·144554	·9816272	0
′	COSINE.	COTANG.	TANG.	SINE.	′

Deg. 79.

NATURAL SINES AND TANGENTS TO A RADIUS 1.

11 Deg.

′	SINE.	TANG.	COTANG.	COSINE.	′
0	·1908090	·194380	5·144554	·9816272	60
1	·1910945	·194682	5·136576	·9815716	59
2	·1913801	·194984	5·128622	·9815160	58
3	·1916656	·195286	5·120692	·9814603	57
4	·1919510	·195588	5·112785	·9814045	56
5	·1922365	·195890	5·104902	·9813486	55
6	·1925220	·196192	5·097042	·9812927	54
7	·1928074	·196494	5·089206	·9812366	53
8	·1930928	·196796	5·081392	·9811805	52
9	·1933782	·197098	5·073602	·9811243	51
10	·1936636	·197400	5·065835	·9810680	50
11	·1939490	·197703	5·058090	·9810116	49
12	·1942344	·198005	5·050369	·9809552	48
13	·1945197	·198307	5·042670	·9808986	47
14	·1948050	·198610	5·034993	·9808420	46
15	·1950903	·198912	5·027339	·9807853	45
16	·1953756	·199214	5·019707	·9807285	44
17	·1956609	·199517	5·012098	·9806716	43
18	·1959461	·199819	5·004511	·9806147	42
19	·1962314	·200122	4·996945	·9805576	41
20	·1965166	·200424	4·989402	·9805005	40
21	·1968018	·200727	4·981881	·9804433	39
22	·1970870	·201030	4·974381	·9803860	38
23	·1973722	·201332	4·966903	·9803286	37
24	·1976573	·201635	4·959447	·9802712	36
25	·1979425	·201938	4·952012	·9802136	35
26	·1982276	·202240	4·944599	·9801560	34
27	·1985127	·202543	4·937206	·9800983	33
28	·1987978	·202846	4·929835	·9800405	32
29	·1990829	·203149	4·922485	·9799827	31
30	·1993679	·203452	4·915157	·9799247	30
′	COSINE.	COTANG.	TANG.	SINE.	′

Deg. 78.

11 Deg.

′	SINE.	TANG.	COTANG.	COSINE.	′
31	·1996530	·203755	4·907849	·9798667	29
32	·1999380	·204058	4·900562	·9798086	28
33	·2002230	·204361	4·893295	·9797504	27
34	·2005080	·204664	4·886049	·9796921	26
35	·2007930	·204967	4·878824	·9796337	25
36	·2010779	·205270	4·871620	·9795752	24
37	·2013629	·205573	4·864435	·9795167	23
38	·2016478	·205876	4·857271	·9794581	22
39	·2019327	·206180	4·850128	·9793994	21
40	·2022176	·206483	4·843004	·9793406	20
41	·2025024	·206786	4·835901	·9792818	19
42	·2027873	·207090	4·828817	·9792228	18
43	·2030721	·207393	4·821753	·9791638	17
44	·2033569	·207696	4·814709	·9791047	16
45	·2036418	·208000	4·807685	·9790455	15
46	·2039265	·208303	4·800680	·9789862	14
47	·2042113	·208607	4·793695	·9789268	13
48	·2044961	·208910	4·786730	·9788674	12
49	·2047808	·209214	4·779783	·9788079	11
50	·2050655	·209518	4·772856	·9787483	10
51	·2053502	·209821	4·765949	·9786886	9
52	·2056349	·210125	4·759060	·9786288	8
53	·2059195	·210429	4·752190	·9785689	7
54	·2062042	·210733	4·745340	·9785090	6
55	·2064888	·211036	4·738508	·9784490	5
56	·2067734	·211340	4·731695	·9783889	4
57	·2070580	·211644	4·724901	·9783287	3
58	·2073426	·211948	4·718125	·9782684	2
59	·2076272	·212252	4·711368	·9782080	1
60	·2079117	·212556	4·704630	·9781476	0
′	COSINE.	COTANG.	TANG.	SINE.	′

Deg. 78.

NATURAL SINES AND TANGENTS TO A RADIUS 1.

12 Deg. | 12 Deg.

′	SINE.	TANG.	COTANG.	COSINE.	′	′	SINE.	TANG.	COTANG.	COSINE.	′
0	·2079117	·212556	4·704630	·9781476	60	31	·2167236	·221999	4·504507	·9762330	29
1	·2081962	·212860	4·697910	·9780871	59	32	·2170076	·222305	4·498322	·9761699	28
2	·2084807	·213164	4·691208	·9780265	58	33	·2172915	·222610	4·492153	·9761067	27
3	·2087652	·213468	4·684524	·9779658	57	34	·2175754	·222915	4·486000	·9760435	26
4	·2090497	·213773	4·677859	·9779050	56	35	·2178593	·223221	4·479863	·9759802	25
5	·2093341	·214077	4·671212	·9778441	55	36	·2181432	·223526	4·473742	·9759168	24
6	·2096186	·214381	4·664583	·9777832	54	37	·2184271	·223831	4·467637	·9758533	23
7	·2099030	·214685	4·657972	·9777222	53	38	·2187110	·224137	4·461548	·9757897	22
8	·2101874	·214990	4·651378	·9776611	52	39	·2189948	·224442	4·455475	·9757260	21
9	·2104718	·215294	4·644803	·9775999	51	40	·2192786	·224748	4·449418	·9756623	20
10	·2107561	·215598	4·638245	·9775386	50	41	·2195624	·225054	4·443376	·9755985	19
11	·2110405	·215903	4·631705	·9774773	49	42	·2198462	·225359	4·437350	·9755345	18
12	·2113248	·216207	4·625183	·9774159	48	43	·2201300	·225665	4·431339	·9754706	17
13	·2116091	·216512	4·618678	·9773544	47	44	·2204137	·225971	4·425343	·9754065	16
14	·2118934	·216816	4·612190	·9772928	46	45	·2206974	·226276	4·419364	·9753423	15
15	·2121777	·217121	4·605720	·9772311	45	46	·2209811	·226582	4·413399	·9752781	14
16	·2124619	·217425	4·599268	·9771693	44	47	·2212648	·226888	4·407450	·9752138	13
17	·2127462	·217730	4·592832	·9771075	43	48	·2215485	·227194	4·401516	·9751494	12
18	·2130304	·218035	4·586414	·9770456	42	49	·2218321	·227500	4·395597	·9750849	11
19	·2133146	·218340	4·580012	·9769836	41	50	·2221158	·227806	4·389694	·9750203	10
20	·2135988	·218644	4·573628	·9769215	40	51	·2223994	·228112	4·383805	·9749556	9
21	·2138829	·218949	4·567261	·9768593	39	52	·2226830	·228418	4·377931	·9748909	8
22	·2141671	·219254	4·560911	·9767970	38	53	·2229666	·228724	4·372073	·9748261	7
23	·2144512	·219559	4·554577	·9767347	37	54	·2232501	·229030	4·366229	·9747612	6
24	·2147353	·219864	4·548260	·9766723	36	55	·2235337	·229336	4·360400	·9746962	5
25	·2150194	·220169	4·541960	·9766098	35	56	·2238172	·229642	4·354586	·9746311	4
26	·2153035	·220474	4·535677	·9765472	34	57	·2241007	·229949	4·348786	·9745660	3
27	·2155876	·220779	4·529410	·9764845	33	58	·2243842	·230255	4·343001	·9745008	2
28	·2158716	·221084	4·523160	·9764217	32	59	·2246676	·230561	4·337231	·9744355	1
29	·2161556	·221389	4·516926	·9763589	31	60	·2249511	·230868	4·331475	·9743701	0
30	·2164396	·221694	4·510708	·9762960	30						
′	COSINE.	COTANG.	TANG.	SINE.	′	′	COSINE.	COTANG.	TANG.	SINE.	′

Deg. 77 | Deg. 77.

NATURAL SINES AND TANGENTS TO A RADIUS 1.

13 Deg. | 13 Deg.

′	SINE.	TANG.	COTANG.	COSINE.	′	′	SINE.	TANG.	COTANG.	COSINE.	′
0	·2249511	·230868	4·331475	·9743701	60	31	·2337282	·240386	4·159968	·9723020	29
1	·2252345	·231174	4·325734	·9743046	59	32	·2340110	·240694	4·154650	·9722339	28
2	·2255179	·231481	4·320007	·9742390	58	33	·2342938	·241001	4·149344	·9721658	27
3	·2258013	·231787	4·314295	·9741734	57	34	·2345766	·241309	4·144051	·9720976	26
4	·2260846	·232094	4·308597	·9741077	56	35	·2348594	·241617	4·138771	·9720294	25
5	·2263680	·232400	4·302913	·9740419	55	36	·2351421	·241925	4·133504	·9719610	24
6	·2266513	·232707	4·297244	·9739760	54	37	·2354248	·242233	4·128249	·9718926	23
7	·2269346	·233014	4·291588	·9739100	53	38	·2357075	·242541	4·123007	·9718240	22
8	·2272179	·233320	4·285947	·9738439	52	39	·2359902	·242849	4·117778	·9717554	21
9	·2275012	·233627	4·280319	·9737778	51	40	·2362729	·243157	4·112561	·9716867	20
10	·2277844	·233934	4·274706	·9737116	50	41	·2365555	·243465	4·107356	·9716180	19
11	·2280677	·234241	4·269107	·9736453	49	42	·2368381	·243773	4·102164	·9715491	18
12	·2283509	·234547	4·263521	·9735789	48	43	·2371207	·244081	4·096985	·9714802	17
13	·2286341	·234854	4·257950	·9735124	47	44	·2374033	·244390	4·091817	·9714112	16
14	·2289172	·235161	4·252392	·9734458	46	45	·2376859	·244698	4·086662	·9713421	15
15	·2292004	·235468	4·246848	·9733792	45	46	·2379684	·245006	4·081519	·9712729	14
16	·2294835	·235775	4·241317	·9733125	44	47	·2382510	·245315	4·076389	·9712036	13
17	·2297666	·236082	4·235800	·9732457	43	48	·2385335	·245623	4·071270	·9711343	12
18	·2300497	·236390	4·230297	·9731789	42	49	·2388159	·245932	4·066164	·9710649	11
19	·2303328	·236697	4·224808	·9731119	41	50	·2390984	·246240	4·061070	·9709953	10
20	·2306159	·237004	4·219331	·9730449	40	51	·2393808	·246549	4·055987	·9709258	9
21	·2308989	·237311	4·213869	·9729777	39	52	·2396633	·246857	4·050917	·9708561	8
22	·2311819	·237618	4·208419	·9729105	38	53	·2399457	·247166	4·045859	·9707863	7
23	·2314649	·237926	4·202983	·9728432	37	54	·2402280	·247475	4·040812	·9707165	6
24	·2317479	·238233	4·197560	·9727759	36	55	·2405104	·247783	4·035777	·9706466	5
25	·2320309	·238541	4·192151	·9727084	35	56	·2407927	·248092	4·030755	·9705766	4
26	·2323138	·238848	4·186754	·9726409	34	57	·2410751	·248401	4·025744	·9705065	3
27	·2325967	·239156	4·181371	·9725733	33	58	·2413574	·248710	4·020744	·9704363	2
28	·2328796	·239463	4·176001	·9725056	32	59	·2416396	·249019	4·015757	·9703660	1
29	·2331625	·239771	4·170644	·9724378	31	60	·2419219	·249328	4·010780	·9702957	0
30	·2334454	·240078	4·165299	·9723699	30						
′	COSINE.	COTANG.	TANG.	SINE.	′	′	COSINE.	COTANG.	TANG.	SINE.	′

Deg. 76. | Deg. 76.

NATURAL SINES AND TANGENTS TO A RADIUS 1.

14 Deg.

′	SINE.	TANG.	COTANG.	COSINE.	′
0	·2419219	·249328	4·010780	·9702957	60
1	·2422041	·249637	4·005816	·9702253	59
2	·2424863	·249946	4·000863	·9701548	58
3	·2427685	·250255	3·995922	·9700842	57
4	·2430507	·250564	3·990992	·9700135	56
5	·2433329	·250873	3·986073	·9699428	55
6	·2436150	·251182	3·981166	·9698720	54
7	·2438971	·251491	3·976271	·9698011	53
8	·2441792	·251801	3·971386	·9697301	52
9	·2444613	·252110	3·966513	·9696591	51
10	·2447433	·252420	3·961651	·9695879	50
11	·2450254	·252729	3·956801	·9695167	49
12	·2453074	·253038	3·951961	·9694453	48
13	·2455894	·253348	3·947133	·9693740	47
14	·2458713	·253658	3·942315	·9693025	46
15	·2461533	·253967	3·937509	·9692309	45
16	·2464352	·254277	3·932714	·9691593	44
17	·2467171	·254587	3·927929	·9690875	43
18	·2469990	·254896	3·923156	·9690157	42
19	·2472809	·255206	3·918393	·9689438	41
20	·2475627	·255516	3·913642	·9688719	40
21	·2478445	·255826	3·908901	·9687998	39
22	·2481263	·256136	3·904171	·9687277	38
23	·2484081	·256446	3·899451	·9686555	37
24	·2486899	·256756	3·894742	·9685832	36
25	·2489716	·257066	3·890044	·9685108	35
26	·2492533	·257376	3·885357	·9684383	34
27	·2495350	·257686	3·880680	·9683658	33
28	·2498167	·257997	3·876014	·9682931	32
29	·2500984	·258307	3·871358	·9682204	31
30	·2503800	·258617	3·866713	·9681476	30
′	COSINE.	COTANG.	TANG.	SINE.	′

Deg. 75.

14 Deg.

′	SINE.	TANG.	COTANG.	COSINE.	′
31	·2506616	·258928	3·862078	·9680748	29
32	·2509432	·259238	3·857453	·9680018	28
33	·2512248	·259548	3·852839	·9679288	27
34	·2515063	·259859	3·848235	·9678557	26
35	·2517879	·260169	3·843642	·9677825	25
36	·2520694	·260480	3·839059	·9677092	24
37	·2523508	·260791	3·834486	·9676358	23
38	·2526323	·261101	3·829923	·9675624	22
39	·2529137	·261412	3·825370	·9674888	21
40	·2531952	·261723	3·820828	·9674152	20
41	·2534766	·262034	3·816295	·9673415	19
42	·2537579	·262345	3·811773	·9672678	18
43	·2540393	·262656	3·807260	·9671939	17
44	·2543206	·262967	3·802758	·9671200	16
45	·2546019	·263278	3·798266	·9670459	15
46	·2548832	·263589	3·793783	·9669718	14
47	·2551645	·263900	3·789310	·9668977	13
48	·2554458	·264211	3·784848	·9668234	12
49	·2557270	·264522	3·780395	·9667490	11
50	·2560082	·264833	3·775951	·9666746	10
51	·2562894	·265145	3·771518	·9666001	9
52	·2565705	·265456	3·767094	·9665255	8
53	·2568517	·265768	3·762680	·9664508	7
54	·2571328	·266079	3·758276	·9663761	6
55	·2574139	·266390	3·753881	·9663012	5
56	·2576950	·266702	3·749496	·9662263	4
57	·2579760	·267014	3·745120	·9661513	3
58	·2582570	·267325	3·740754	·9660762	2
59	·2585381	·267637	3·736398	·9660011	1
60	·2588190	·267949	3·732050	·9659258	0
′	COSINE.	COTANG.	TANG.	SINE.	′

Deg. 75.

NATURAL SINES AND TANGENTS TO A RADIUS 1.

15 DEG. 15 DEG

′	SINE.	TANG.	COTANG.	COSINE.	′	′	SINE.	TANG.	COTANG.	COSINE.	′
0	·2588190	·267949	3·732050	·9659258	60	31	·2675187	·277637	3·601814	·9635527	29
1	·2591000	·268261	3·727713	·9658505	59	32	·2677989	·277951	3·597754	·9634748	28
2	·2593810	·268572	3·723384	·9657751	58	33	·2680792	·278264	3·593702	·9633969	27
3	·2596619	·268884	3·719065	·9656996	57	34	·2683594	·278578	3·589659	·9633189	26
4	·2599428	·269196	3·714756	·9656240	56	35	·2686396	·278891	3·585624	·9632408	25
5	·2602237	·269508	3·710455	·9655484	55	36	·2689198	·279205	3·581597	·9631626	24
6	·2605045	·269820	3·706164	·9654726	54	37	·2692000	·279518	3·577579	·9630843	23
7	·2607853	·270132	3·701883	·9653968	53	38	·2694801	·279832	3·573569	·9630060	22
8	·2610662	·270444	3·697610	·9653209	52	39	·2697602	·280145	3·569568	·9629275	21
9	·2613469	·270757	3·693346	·9652449	51	40	·2700403	·280459	3·565574	·9628490	20
10	·2616277	·271069	3·689092	·9651689	50	41	·2703204	·280773	3·561590	·9627704	19
11	·2619085	·271381	3·684847	·9650927	49	42	·2706004	·281087	3·557613	·9626917	18
12	·2621892	·271694	3·680611	·9650165	48	43	·2708805	·281401	3·553644	·9626130	17
13	·2624699	·272006	3·676384	·9649402	47	44	·2711605	·281715	3·549684	·9625342	16
14	·2627506	·272318	3·672166	·9648638	46	45	·2714404	·282029	3·545732	·9624552	15
15	·2630312	·272631	3·667957	·9647873	45	46	·2717204	·282343	3·541788	·9623762	14
16	·2633118	·272943	3·663757	·9647108	44	47	·2720003	·282657	3·537852	·9622972	13
17	·2635925	·273256	3·659566	·9646341	43	48	·2722802	·282971	3·533925	·9622180	12
18	·2638730	·273569	3·655384	·9645574	42	49	·2725601	·283285	3·530005	·9621387	11
19	·2641536	·273881	3·651211	·9644806	41	50	·2728400	·283599	3·526093	·9620594	10
20	·2644342	·274194	3·647046	·9644037	40	51	·2731198	·283914	3·522190	·9619800	9
21	·2647147	·274507	3·642891	·9643268	39	52	·2733997	·284228	3·518294	·9619005	8
22	·2649952	·274820	3·638744	·9642497	38	53	·2736794	·284543	3·514407	·9618210	7
23	·2652757	·275133	3·634606	·9641726	37	54	·2739592	·284857	3·510527	·9617413	6
24	·2655561	·275445	3·630477	·9640954	36	55	·2742390	·285172	3·506655	·9616616	5
25	·2658366	·275758	3·626356	·9640181	35	56	·2745187	·285486	3·502791	·9615818	4
26	·2661170	·276071	3·622244	·9639407	34	57	·2747984	·285801	3·498935	·9615019	3
27	·2663973	·276385	3·618141	·9638633	33	58	·2750781	·286115	3·495087	·9614219	2
28	·2666777	·276698	3·614046	·9637858	32	59	·2753577	·286430	3·491247	·9613418	1
29	·2669581	·277011	3·609960	·9637081	31	60	·2756374	·286745	3·487414	·9612617	0
30	·2672384	·277324	3·605883	·9636305	30						
′	COSINE.	COTANG.	TANG.	SINE.	′	′	COSINE.	COTANG.	TANG.	SINE.	′

DEG. 74. DEG. 74.

NATURAL SINES AND TANGENTS TO A RADIUS 1.

16 Deg. | 16 Deg.

′	SINE.	TANG.	COTANG.	COSINE.	′	′	SINE.	TANG.	COTANG.	COSINE.	′
0	·2756374	·286745	3·487414	·9612617	60	31	·2842942	·296529	3·372340	·9587371	29
1	·2759170	·287060	3·483589	·9611815	59	32	·2845731	·296846	3·368745	·9586543	28
2	·2761965	·287375	3·479772	·9611012	58	33	·2848520	·297163	3·365156	·9585715	27
3	·2764761	·287690	3·475963	·9610208	57	34	·2851308	·297479	3·361575	·9584886	26
4	·2767556	·288005	3·472161	·9609403	56	35	·2854096	·297796	3·358000	·9584056	25
5	·2770352	·288320	3·468367	·9608598	55	36	·2856884	·298112	3·354433	·9583226	24
6	·2773147	·288635	3·464581	·9607792	54	37	·2859671	·298429	3·350872	·9582394	23
7	·2775941	·288950	3·460802	·9606984	53	38	·2862458	·298746	3·347319	·9581562	22
8	·2778736	·289265	3·457031	·9606177	52	39	·2865246	·299063	3·343772	·9580729	21
9	·2781530	·289580	3·453267	·9605368	51	40	·2868032	·299380	3·340232	·9579895	20
10	·2784324	·289896	3·449512	·9604558	50	41	·2870819	·299697	3·336699	·9579060	19
11	·2787118	·290211	3·445763	·9603748	49	42	·2873605	·300014	3·333173	·9578225	18
12	·2789911	·290526	3·442022	·9602937	48	43	·2876391	·300331	3·329654	·9577389	17
13	·2792704	·290842	3·438289	·9602125	47	44	·2879177	·300648	3·326141	·9576552	16
14	·2795497	·291157	3·434563	·9601312	46	45	·2881963	·300965	3·322636	·9575714	15
15	·2798290	·291473	3·430844	·9600499	45	46	·2884748	·301283	3·319137	·9574875	14
16	·2801083	·291789	3·427133	·9599684	44	47	·2887533	·301600	3·315645	·9574035	13
17	·2803875	·292104	3·423429	·9598869	43	48	·2890318	·301917	3·312159	·9573195	12
18	·2806667	·292420	3·419733	·9598053	42	49	·2893103	·302235	3·308681	·9572354	11
19	·2809459	·292736	3·416044	·9597236	41	50	·2895887	·302552	3·305209	·9571512	10
20	·2812251	·293052	3·412362	·9596418	40	51	·2898671	·302870	3·301743	·9570669	9
21	·2815042	·293368	3·408688	·9595600	39	52	·2901455	·303187	3·298285	·9569825	8
22	·2817833	·293683	3·405021	·9594781	38	53	·2904239	·303505	3·294833	·9568981	7
23	·2820624	·293999	3·401361	·9593961	37	54	·2907022	·303823	3·291387	·9568136	6
24	·2823415	·294316	3·397708	·9593140	36	55	·2909805	·304141	3·287948	·9567290	5
25	·2826205	·294632	3·394063	·9592318	35	56	·2912588	·304458	3·284516	·9566443	4
26	·2828995	·294948	3·390424	·9591496	34	57	·2915371	·304776	3·281090	·9565595	3
27	·2831785	·295264	3·386793	·9590672	33	58	·2918153	·305094	3·277671	·9564747	2
28	·2834575	·295580	3·383169	·9589848	32	59	·2920935	305412	3·274258	·9563898	1
29	·2837364	·295897	3·379553	·9589023	31	60	·2923717	305730	3·270852	·9563048	0
30	·2840153	·296213	3·375943	·9588197	30						
′	COSINE.	COTANG.	TANG.	SINE.	′	′	COSINE.	COTANG.	TANG.	SINE.	′

Deg. 73. | Deg. 73.

NATURAL SINES AND TANGENTS TO A RADIUS 1.

17 Deg. | 17 Deg.

′	SINE.	TANG.	COTANG.	COSINE.	′	′	SINE.	TANG.	COTANG.	COSINE.	′
0	·2923717	·305730	3·270852	·9563048	60	31	·3009832	·315618	3·168380	·9536294	29
1	·2926499	·306048	3·267452	·9562197	59	32	·3012606	·315938	3·165172	·9535418	28
2	·2929280	·306367	3·264059	·9561345	58	33	·3015380	·316258	3·161970	·9534542	27
3	·2932061	·306685	3·260672	·9560492	57	34	·3018153	·316578	3·158774	·9533664	26
4	·2934842	·307003	3·257292	·9559639	56	35	·3020926	·316898	3·155584	·9532786	25
5	·2937623	·307321	3·253918	·9558785	55	36	·3023699	·317218	3·152399	·9531907	24
6	·2940403	·307640	3·250550	·9557930	54	37	·3026471	·317538	3·149220	·9531027	23
7	·2943183	·307958	3·247189	·9557074	53	38	·3029244	·317859	3·146047	·9530146	22
8	·2945963	·308277	3·243834	·9556218	52	39	·3032016	·318179	3·142880	·9529264	21
9	·2948743	·308595	3·240486	·9555361	51	40	·3034788	·318499	3·139719	·9528382	20
10	·2951522	·308914	3·237143	·9554502	50	41	·3037559	·318820	3·136563	·9527499	19
11	·2954302	·309233	3·233807	·9553643	49	42	·3040331	·319140	3·133414	·9526615	18
12	·2957081	·309551	3·230478	·9552784	48	43	·3043102	·319461	3·130270	·9525730	17
13	·2959859	·309870	3·227154	·9551923	47	44	·3045872	·319781	3·127131	·9524844	16
14	·2962638	·310189	3·223837	·9551062	46	45	·3048643	·320102	3·123999	·9523958	15
15	·2965416	·310508	3·220526	·9550199	45	46	·3051413	·320423	3·120872	·9523071	14
16	·2968194	·310827	3·217221	·9549336	44	47	·3054183	·320744	3·117750	·9522183	13
17	·2970971	·311146	3·213922	·9548473	43	48	·3056953	·321064	3·114635	·9521294	12
18	·2973749	·311465	3·210630	·9547608	42	49	·3059723	·321385	3·111525	·9520404	11
19	·2976526	·311784	3·207344	·9546743	41	50	·3062492	·321706	3·108421	·9519514	10
20	·2979303	·312103	3·204063	·9545876	40	51	·3065261	·322027	3·105322	·9518623	9
21	·2982079	·312422	3·200789	·9545009	39	52	·3068030	·322348	3·102229	·9517731	8
22	·2984856	·312742	3·197521	·9544141	38	53	·3070798	·322670	3·099141	·9516838	7
23	·2987632	·313061	3·194259	·9543273	37	54	·3073566	·322991	3·096059	·9515944	6
24	·2990408	·313381	3·191003	·9542403	36	55	·3076334	·323312	3·092983	·9515050	5
25	·2993184	·313700	3·187754	·9541533	35	56	·3079102	·323633	3·089912	·9514154	4
26	·2995959	·314020	3·184510	·9540662	34	57	·3081869	·323955	3·086846	·9513258	3
27	·2998734	·314339	3·181272	·9539790	33	58	·3084636	·324276	3·083786	·9512361	2
28	·3001509	·314659	3·178040	·9538917	32	59	·3087403	·324598	3·080732	·9511464	1
29	·3004284	·314979	3·174814	·9538044	31	60	·3090170	·324919	3·077683	·9510565	0
30	·3007058	·315298	3·171594	·9537170	30						
′	COSINE.	COTANG.	TANG.	SINE.	′	′	COSINE.	COTANG.	TANG.	SINE.	′

Deg. 72. | Deg. 72.

NATURAL SINES AND TANGENTS TO A RADIUS 1.

18 Deg.

′	SINE.	TANG.	COTANG.	COSINE.	′
0	·3090170	·324919	3·077683	·9510565	60
1	·3092936	·325241	3·074640	·9509666	59
2	·3095702	·325563	3·071602	·9508766	58
3	·3098468	·325884	3·068569	·9507865	57
4	·3101234	·326206	3·065542	·9506963	56
5	·3103999	·326528	3·062520	·9506061	55
6	·3106764	·326850	3·059503	·9505157	54
7	·3109529	·327172	3·056492	·9504253	53
8	·3112294	·327494	3·053487	·9503348	52
9	·3115058	·327816	3·050486	·9502443	51
10	·3117822	·328138	3·047491	·9501536	50
11	·3120586	·328461	3·044501	·9500629	49
12	·3123349	·328783	3·041517	·9499721	48
13	·3126112	·329105	3·038538	·9498812	47
14	·3128875	·329428	3·035564	·9497902	46
15	·3131638	·329750	3·032595	·9496991	45
16	·3134400	·330073	3·029632	·9496080	44
17	·3137163	·330395	3·026673	·9495168	43
18	·3139925	·330718	3·023720	·9494255	42
19	·3142686	·331041	3·020772	·9493341	41
20	·3145448	·331363	3·017830	·9492426	40
21	·3148209	·331686	3·014892	·9491511	39
22	·3150969	·332009	3·011960	·9490595	38
23	·3153730	·332332	3·009033	·9489678	37
24	·3156490	·332655	3·006110	·9488760	36
25	·3159250	·332978	3·003193	·9487842	35
26	·3162010	·333302	3·000282	·9486922	34
27	·3164770	·333625	2·997375	·9486002	33
28	·3167529	·333948	2·994473	·9485081	32
29	·3170288	·334271	2·991576	·9484159	31
30	·3173047	·334595	2·988685	·9483237	30
′	COSINE.	COTANG.	TANG.	SINE.	′

Deg. 71.

18 Deg.

′	SINE.	TANG.	COTANG.	COSINE.	′
31	·3175805	·334918	2·985798	·9482313	29
32	·3178563	·335242	2·982916	·9481389	28
33	·3181321	·335566	2·980040	·9480464	27
34	·3184079	·335889	2·977168	·9479538	26
35	·3186836	·336213	2·974301	·9478612	25
36	·3189593	·336537	2·971439	·9477684	24
37	·3192350	·336861	2·968583	·9476756	23
38	·3195106	·337185	2·965731	·9475827	22
39	·3197863	·337509	2·962884	·9474897	21
40	·3200619	·337833	2·960042	·9473966	20
41	·3203374	·338157	2·957205	·9473035	19
42	·3206130	·338481	2·954372	·9472103	18
43	·3208885	·338805	2·951545	·9471170	17
44	·3211640	·339129	2·948722	·9470236	16
45	·3214395	·339454	2·945905	·9469301	15
46	·3217149	·339778	2·943092	·9468366	14
47	·3219903	·340103	2·940284	·9467430	13
48	·3222657	·340427	2·937480	·9466493	12
49	·3225411	·340752	2·934682	·9465555	11
50	·3228164	·341077	2·931888	·9464616	10
51	·3230917	·341401	2·929099	·9463677	9
52	·3233670	·341726	2·926315	·9462736	8
53	·3236422	·342051	2·923535	·9461795	7
54	·3239174	·342376	2·920761	·9460854	6
55	·3241926	·342701	2·917990	·9459911	5
56	·3244678	·343026	2·915225	·9458968	4
57	·3247429	·343351	2·912464	·9458023	3
58	·3250180	·343677	2·909708	·9457078	2
59	·3252931	·344002	2·906957	·9456132	1
60	·3255682	·344327	2·904210	·9455186	0
′	COSINE.	COTANG.	TANG.	SINE.	′

Deg. 71.

NATURAL SINES AND TANGENTS TO A RADIUS 1.

19 Deg.

′	SINE.	TANG.	COTANG.	COSINE.	′
0	·3255682	·344327	2·904210	·9455186	60
1	·3258432	·344653	2·901468	·9454238	59
2	·3261182	·344978	2·898731	·9453290	58
3	·3263932	·345304	2·895998	·9452341	57
4	·3266681	·345629	2·893270	·9451391	56
5	·3269430	·345955	2·890546	·9450441	55
6	·3272179	·346281	2·887827	·9449489	54
7	·3274928	·346606	2·885113	·9448537	53
8	·3277676	·346932	2·882403	·9447584	52
9	·3280424	·347258	2·879697	·9446630	51
10	·3283172	·347584	2·876997	·9445675	50
11	·3285919	·347910	2·874300	·9444720	49
12	·3288666	·348236	2·871608	·9443764	48
13	·3291413	·348563	2·868921	·9442807	47
14	·3294160	·348889	2·866238	·9441849	46
15	·3296906	·349215	2·863560	·9440890	45
16	·3299653	·349542	2·860886	·9439931	44
17	·3302398	·349868	2·858216	·9438971	43
18	·3305144	·350195	2·855551	·9438010	42
19	·3307889	·350521	2·852891	·9437048	41
20	·3310634	·350848	2·850234	·9436085	40
21	·3313379	·351175	2·847583	·9435122	39
22	·3316123	·351501	2·844935	·9434157	38
23	·3318867	·351828	2·842292	·9433192	37
24	·3321611	·352155	2·839653	·9432227	36
25	·3324355	·352482	2·837019	·9431260	35
26	·3327098	·352809	2·834389	·9430293	34
27	·3329841	·353136	2·831763	·9429324	33
28	·3332584	·353464	2·829142	·9428355	32
29	·3335326	·353791	2·826525	·9427386	31
30	·3338069	·354118	2·823912	·9426415	30
′	COSINE.	COTANG.	TANG.	SINE.	′

Deg. 70.

19 Deg.

′	SINE.	TANG.	COTANG.	COSINE.	′
31	·3340810	·354446	2·821304	·9425444	29
32	·3343552	·354773	2·818700	·9424471	28
33	·3346293	·355101	2·816100	·9423498	27
34	·3349034	·355428	2·813504	·9422525	26
35	·3351775	·355756	2·810913	·9421550	25
36	·3354516	·356084	2·808326	·9420575	24
37	·3357256	·356411	2·805743	·9419598	23
38	·3359996	·356739	2·803164	·9418621	22
39	·3362735	·357067	2·800590	·9417644	21
40	·3365475	·357395	2·798019	·9416665	20
41	·3368214	·357723	2·795453	·9415686	19
42	·3370953	·358051	2·792891	·9414705	18
43	·3373691	·358380	2·790333	·9413724	17
44	·3376429	·358708	2·787789	·9412743	16
45	·3379167	·359036	2·785230	·9411760	15
46	·3381905	·359365	2·782685	·9410777	14
47	·3384642	·359693	2·780144	·9409793	13
48	·3387379	·360022	2·777606	·9408808	12
49	·3390116	·360350	2·775073	·9407822	11
50	·3392852	·360679	2·772544	·9406835	10
51	·3395589	·361008	2·770019	·9405848	9
52	·3398325	·361337	2·767499	·9404860	8
53	·3401060	·361666	2·764982	·9403871	7
54	·3403796	·361994	2·762469	·9402881	6
55	·3406531	·362324	2·759960	·9401891	5
56	·3409265	·362653	2·757456	·9400899	4
57	·3412000	·362982	2·754955	·9399907	3
58	·3414734	·363311	2·752458	·9398914	2
59	·3417468	·363640	2·749966	·9397921	1
60	·3420201	·363970	2·747477	·9396926	0
′	COSINE.	COTANG.	TANG.	SINE.	′

Deg. 70.

NATURAL SINES AND TANGENTS TO A RADIUS 1.

20 Deg. 20 Deg.

′	SINE.	TANG.	COTANG.	COSINE.	′	′	SINE.	TANG.	COTANG.	COSINE.	′
0	·3420201	·363970	2·747477	·9396926	60	31	·3504798	·374216	2·672251	·9365703	29
1	·3422935	·364299	2·744992	·9395931	59	32	·3507523	·374547	2·669885	·9364683	28
2	·3425668	·364629	2·742512	·9394935	58	33	·3510246	·374879	2·667522	·9363662	27
3	·3428400	·364958	2·740035	·9393938	57	34	·3512970	·375211	2·665163	·9362641	26
4	·3431133	·365288	2·737562	·9392940	56	35	·3515693	·375543	2·662808	·9361618	25
5	·3433865	·365618	2·735093	·9391942	55	36	·3518416	·375875	2·660456	·9360595	24
6	·3436597	·365948	2·732628	·9390943	54	37	·3521139	·376207	2·658108	·9359571	23
7	·3439329	·366277	2·730167	·9389943	53	38	·3523862	·376539	2·655764	·9358547	22
8	·3442060	·366607	2·727710	·9388942	52	39	3526584	·376871	2·653423	·9357521	21
9	·3444791	·366937	2·725256	·9387940	51	40	3529306	·377203	2·651086	·9356495	20
10	·3447521	·367268	2·722807	·9386938	50	41	·3532027	·377536	2·648753	·9355468	19
11	·3450252	·367598	2·720362	·9385934	49	42	·3534748	·377868	2·646423	·9354440	18
12	·3452982	·367928	2·717920	·9384930	48	43	·3537469	·378201	2·644096	·9353412	17
13	·3455712	·368258	2·715482	·9383925	47	44	·3540190	·378533	2·641774	·9352382	16
14	·3458441	·368589	2·713048	·9382920	46	45	·3542910	·378866	2·639454	·9351352	15
15	·3461171	·368919	2·710618	·9381913	45	46	·3545630	·379198	2·637139	·9350321	14
16	·3463900	·369250	2·708192	·9380906	44	47	·3548350	·379531	2·634827	·9349289	13
17	·3466628	·369580	2·705769	·9379898	43	48	·3551070	·379864	2·632518	·9348257	12
18	·3469357	·369911	2·703351	·9378889	42	49	·3553789	·380197	2·630213	·9347223	11
19	·3472085	·370242	2·700936	·9377880	41	50	·3556508	·380530	2·627912	·9346189	10
20	·3474812	·370572	2·698525	·9376869	40	51	·3559226	·380863	2·625614	·9345154	9
21	·3477540	·370903	2·696118	·9375858	39	52	·3561944	·381196	2·623319	·9344119	8
22	·3480267	·371234	2·693714	·9374846	38	53	·3564662	·381529	2·621028	·9343082	7
23	·3482994	·371565	2·691314	·9373833	37	54	·3567380	·381862	2·618741	·9342045	6
24	·3485720	·371896	2·688919	·9372820	36	55	·3570097	·382196	2·616457	·9341007	5
25	·3488447	·372227	2·686526	·9371806	35	56	·3572814	·382529	2·614176	·9339968	4
26	·3491173	·372559	2·684138	·9370790	34	57	·3575531	·382863	2·611899	·9338928	3
27	·3493898	·372890	2·681753	·9369774	33	58	·3578248	·383196	2·609625	·9337888	2
28	·3496624	·373221	2·679372	·9368758	32	59	·3580964	·383530	2·607355	·9336846	1
29	·3499349	·373553	2·676995	·9367740	31	60	·3583679	·383864	2·605089	·9335804	0
30	·3502074	·373884	2·674621	·9366722	30						
′	COSINE.	COTANG.	TANG.	SINE.	′	′	COSINE.	COTANG.	TANG.	SINE.	′

Deg. 69. Deg. 69.

NATURAL SINES AND TANGENTS TO A RADIUS 1.

21 Deg.

′	SINE.	TANG.	COTANG.	COSINE.	′
0	·3583679	·383864	2·605089	·9335804	60
1	·3586395	·384197	2·602825	·9334761	59
2	·3589110	·384531	2·600565	·9333718	58
3	·3591825	·384865	2·598309	·9332673	57
4	·3594540	·385199	2·596056	·9331628	56
5	·3597254	·385533	2·593806	·9330582	55
6	·3599968	·385867	2·591560	·9329535	54
7	·3602682	·386202	2·589317	·9328488	53
8	·3605395	·386536	2·587078	·9327439	52
9	·3608102	·386870	2·584842	·9326390	51
10	·3610821	·387205	2·582609	·9325340	50
11	·3613534	·387539	2·580380	·9324290	49
12	·3616246	·387874	2·578153	·9323238	48
13	·3618958	·388209	2·575931	·9322186	47
14	·3621699	·388543	2·573711	·9321133	46
15	·3624380	·388878	2·571495	·9320079	45
16	·3627091	·389213	2·569283	·9319024	44
17	·3629802	·389548	2·567073	·9317969	43
18	·3632512	·389883	2·564867	·9316912	42
19	·3635222	·390218	2·562664	·9315855	41
20	·3637932	·390554	2·560464	·9314797	40
21	·3640641	·390889	2·558268	·9313739	39
22	·3643351	·391224	2·556075	·9312679	38
23	·3646059	·391560	2·553885	·9311619	37
24	·3648768	·391895	2·551699	·9310558	36
25	·3651476	·392231	2·549516	·9309496	35
26	·3654184	·392567	2·547335	·9308434	34
27	·3656891	·392902	2·545159	·9307370	33
28	·3659599	·393238	2·542985	·9306306	32
29	·3662306	·393574	2·540815	·9305241	31
30	·3665012	·393910	2·538647	·9304176	30
′	COSINE.	COTANG.	TANG.	SINE.	′

Deg. 68.

21 Deg.

′	SINE.	TANG.	COTANG.	COSINE.	′
31	·3667719	·394246	2·536483	·9303109	29
32	·3670425	·394582	2·534323	·9302042	28
33	·3673130	·394918	2·532165	·9300974	27
34	·3675836	·395255	2·530011	·9299905	26
35	·3678541	·395591	2·527859	·9298835	25
36	·3681246	·395928	2·525711	·9297765	24
37	·3683950	·396264	2·523566	·9296694	23
38	·3686654	·396601	2·521424	·9295622	22
39	·3689358	·396937	2·519286	·9294549	21
40	·3692061	·397274	2·517150	·9293475	20
41	·3694765	·397611	2·515018	·9292401	19
42	·3697468	·397948	2·512889	·9291326	18
43	·3700170	·398285	2·510762	·9290250	17
44	·3702872	·398622	2·508639	·9289173	16
45	·3705574	·398959	2·506519	·9288096	15
46	·3708276	·399296	2·504402	·9287017	14
47	·3710977	·399634	2·502289	·9285938	13
48	·3713678	·399971	2·500178	·9284858	12
49	·3716379	·400308	2·498070	·9283778	11
50	·3719079	·400646	2·495966	·9282696	10
51	·3721780	·400984	2.493864	·9281614	9
52	·3724479	·401321	2·491766	·9280531	8
53	·3727179	·401659	2·489670	·9279447	7
54	·3729878	·401997	2·487578	·9278363	6
55	·3732577	·402335	2·485488	·9277277	5
56	·3735275	·402673	2·483402	·9276191	4
57	·3737973	·403011	2·481319	·9275104	3
58	·3740671	·403349	2·479238	·9274016	2
59	·3743369	·403687	2·477161	·9272928	1
60	·3746066	·404026	2·475086	·9271839	0
′	COSINE.	COTANG.	TANG.	SINE.	′

Deg. 68.

NATURAL SINES AND TANGENTS TO A RADIUS 1.

22 Deg. 22 Deg.

′	SINE.	TANG.	COTANG.	COSINE.	′	′	SINE.	TANG.	COTANG.	COSINE.	′
0	·3746066	·404026	2·475086	·9271839	60	31	·3829522	·414554	2·412228	·9237682	29
1	·3748763	·404364	2·473015	·9270748	59	32	·3832209	·414895	2·410246	·9236567	28
2	·3751459	·404703	2·470947	·9269658	58	33	·3834895	·415236	2·408267	·9235452	27
3	·3754156	·405041	2·468881	·9268566	57	34	·3837582	·415577	2·406290	·9234336	26
4	·3756852	·405380	2·466819	·9267474	56	35	·3840268	·415918	2·404316	·9233220	25
5	·3759547	·405719	2·464759	·9266380	55	36	·3842953	·416259	2·402345	·9232102	24
6	·3762243	·406057	2·462703	·9265286	54	37	·3845639	·416601	2·400377	·9230984	23
7	·3764938	·406396	2·460649	·9264192	53	38	·3848324	·416942	2·398411	·9229865	22
8	·3767632	·406735	2·458598	·9263096	52	39	·3851008	·417284	2·396449	·9228745	21
9	·3770327	·407074	2·456551	·9262000	51	40	·3853693	·417625	2·394488	·9227624	20
10	·3773021	·407413	2·454506	·9260902	50	41	·3856377	·417967	2·392531	·9226503	19
11	·3775714	·407753	2·452464	·9259805	49	42	·3859060	·418309	2·390576	·9225381	18
12	·3778408	·408092	2·450425	·9258706	48	43	·3861744	·418650	2·388625	·9224258	17
13	·3781101	·408431	2·448389	·9257606	47	44	·3864427	·418992	2·386675	·9223134	16
14	·3783794	·408771	2·446355	·9256506	46	45	·3867110	·419334	2·384729	·9222010	15
15	·3786486	·409110	2·444325	·9255405	45	46	·3869792	·419676	2·382785	·9220884	14
16	·3789178	·409450	2·442298	·9254303	44	47	·3872474	·420019	2·380844	·9219758	13
17	·3791870	·409790	2·440273	·9253201	43	48	·3875156	·420361	2·378906	·9218632	12
18	·3794562	·410129	2·438251	·9252097	42	49	·3877837	·420703	2·376970	·9217504	11
19	·3797253	·410469	2·436233	·9250993	41	50	·3880518	·421046	2·375037	·9216375	10
20	·3799944	·410809	2·434217	·9249888	40	51	·3883199	·421388	2·373106	·9215246	9
21	·3802634	·411149	2·432204	·9248782	39	52	·3885880	·421731	2·371179	·9214116	8
22	·3805324	·411489	2·430193	·9247676	38	53	·3888560	·422073	2·369254	·9212986	7
23	·3808014	·411830	2·428186	·9246568	37	54	·3891240	·422416	2·367331	·9211854	6
24	·3810704	·412170	2·426181	·9245460	36	55	·3893919	·422759	2·365411	·9210722	5
25	·3813393	·412510	2·424180	·9244351	35	56	·3896598	·423102	2·363494	·9209589	4
26	·3816082	·412851	2·422181	·9243242	34	57	·3899277	·423445	2·361580	·9208455	3
27	·3818770	·413191	2·420185	·9242131	33	58	·3901955	·423788	2·359668	·9207320	2
28	·3821459	·413532	2·418191	·9241020	32	59	·3904633	·424131	2·357759	·9206185	1
29	·3824147	·413872	2·416201	·9239908	31	60	·3907311	·424474	2·355852	·9205049	0
30	·3826834	·414213	2·414213	·9238795	30						
′	COSINE.	COTANG.	TANG.	SINE.	′	′	COSINE.	COTANG.	TANG.	SINE.	′

Deg. 67. Deg. 67.

NATURAL SINES AND TANGENTS TO A RADIUS 1.

23 Deg. | 23 Deg

′	SINE.	TANG.	COTANG.	COSINE.	′	′	SINE.	TANG.	COTANG.	COSINE.	′
0	·3907311	·424474	2·355852	·9205049	60	31	·3990158	·435158	2·298014	·9169440	29
1	·3909989	·424818	2·353948	·9203912	59	32	·3992825	·435504	2·296188	·9168279	28
2	·3912660	·425161	2·352046	·9202774	58	33	·3995492	·435850	2·294365	·9167118	27
3	·3915343	·425505	2·350148	·9201635	57	34	·3998158	·436196	2·292544	·9165955	26
4	·3918019	·425848	2·348251	·9200496	56	35	·4000825	·436542	2·290725	·9164791	25
5	·3920695	·426192	2·346358	·9199356	55	36	·4003490	·436889	2·288909	·9163627	24
6	·3923371	·426536	2·344467	·9198215	54	37	·4006156	·437235	2·287095	·9162462	23
7	·3926047	·426880	2·342578	·9197073	53	38	·4008821	·437582	2·285284	·9161297	22
8	·3928722	·427223	2·340692	·9195931	52	39	·4011486	·437928	2·283475	·9160130	21
9	·3931397	·427568	2·338809	·9194788	51	40	·4014150	·438275	2·281669	·9158963	20
10	·3934071	·427912	2·336928	·9193644	50	41	·4016814	·438622	2·279865	·9157795	19
11	·3936745	·428256	2·335050	·9192499	49	42	·4019478	·438969	2·278063	·9156626	18
12	·3939419	·428600	2·333174	·9191353	48	43	·4022141	·439316	2·276264	·9155456	17
13	·3942093	·428944	2·331301	·9190207	47	44	·4024804	·439663	2·274467	·9154286	16
14	·3944765	·429289	2·329431	·9189060	46	45	·4027467	·440010	2·272672	·9153115	15
15	·3947439	·429633	2·327563	·9187912	45	46	·4030129	·440357	2·270880	·9151943	14
16	·3950111	·429978	2·325697	·9186763	44	47	·4032791	·440705	2·269090	·9150770	13
17	·3952783	·430323	2·323834	·9185614	43	48	·4035453	·441052	2·267303	·9149597	12
18	·3955455	·430668	2·321974	·9184464	42	49	·4038114	·441400	2·265518	·9148422	11
19	·3958127	·431012	2·320116	·9183313	41	50	·4040775	·441747	2·263735	·9147247	10
20	·3960798	·431357	2·318260	·9182161	40	51	·4043436	·442095	2·261955	·9146072	9
21	·3963468	·431703	2·316407	·9181009	39	52	·4046096	·442443	2·260177	·9144895	8
22	·3966139	·432048	2·314557	·9179855	38	53	·4048756	·442791	2·258401	·9143718	7
23	·3968809	·432393	2·312709	·9178701	37	54	·4051416	·443139	2·256628	·9142540	6
24	·3971479	·432738	2·310863	·9177546	36	55	·4054075	·443487	2·254857	·9141361	5
25	·3974148	·433084	2·309020	·9176391	35	56	·4056734	·443835	2·253088	·9140181	4
26	·3976818	·433429	2·307180	·9175234	34	57	·4059393	·444183	2·251322	·9139001	3
27	·3979486	·433775	2·305342	·9174077	33	58	·4062051	·444531	2·249558	·9137819	2
28	·3982155	·434120	2·303506	·9172919	32	59	·4064709	·444880	2·247796	·9136637	1
29	·3984823	·434466	2·301673	·9171760	31	60	·4067366	·445228	2·246036	·9135455	0
30	·3987491	·434812	2·299842	·9170601	30						
′	COSINE.	COTANG.	TANG.	SINE.	′	′	COSINE.	COTANG.	TANG.	SINE.	′

Deg. 66. | Deg. 66

NATURAL SINES AND TANGENTS TO A RADIUS 1.

24 Deg. 24 Deg.

′	SINE.	TANG.	COTANG.	COSINE.	′	′	SINE.	TANG.	COTANG.	COSINE.	′
0	·4067366	445228	2·246036	·9135455	60	31	·4149579	·456077	2·192609	·9098406	29
1	·4070024	·445577	2·244279	·9134271	59	32	·4152226	·456429	2·190921	·9097199	28
2	·4072681	·445926	2·242524	·9133087	58	33	·4154872	·456780	2·189234	·9095990	27
3	·4075337	·446274	2·240772	·9131902	57	34	·4157517	·457132	2·187551	·9094781	26
4	·4077993	·446623	2·239021	·9130716	56	35	·4160163	·457483	2·185869	·9093572	25
5	·4080649	·446972	2·237273	·9129529	55	36	·4162808	·457835	2·184189	·9092361	24
6	·4083305	·447321	2·235528	·9128342	54	37	·4165453	·458187	2·182511	·9091150	23
7	·4085960	·447670	2·233784	·9127154	53	38	·4168097	·458539	2·180836	·9089938	22
8	·4088615	·448020	2·232043	·9125965	52	39	·4170741	·458891	2·179163	·9088725	21
9	·4091269	·448369	2·230304	·9124775	51	40	·4173385	·459243	2·177492	·9087511	20
10	·4093923	·448718	2·228567	·9123584	50	41	·4176028	·459596	2·175822	·9086297	19
11	·4096577	·449068	2·226833	·9122393	49	42	·4178671	·459948	2·174155	·9085082	18
12	·4099230	·449417	2·225100	·9121201	48	43	·4181313	·460301	2·172491	·9083866	17
13	·4101883	·449767	2·223370	·9120008	47	44	·4183956	·460653	2·170828	·9082649	16
14	·4104536	·450117	2·221643	·9118815	46	45	·4186597	·461006	2·169167	·9081432	15
15	·4107189	·450467	2·219917	·9117620	45	46	·4189239	·461359	2·167509	·9080214	14
16	·4109841	·450817	2·218194	·9116425	44	47	·4191880	·461711	2·165852	·9078995	13
17	·4112492	·451167	2·216473	·9115229	43	48	·4194521	·462064	2·164198	·9077775	12
18	·4115144	·451517	2·214754	·9114033	42	49	·4197161	·462417	2·162546	·9076554	11
19	·4117795	·451867	2·213037	·9112835	41	50	·4199801	·462771	2·160895	·9075333	10
20	·4120445	·452217	2·211323	·9111637	40	51	·4202441	·463124	2·159247	·9074111	9
21	·4123096	452568	2·209611	·9110438	39	52	·4205080	·463477	2·157601	·9072888	8
22	·4125745	·452918	2·207901	·9109238	38	53	·4207719	·463831	2·155957	·9071665	7
23	·4128395	·453269	2·206193	·9108038	37	54	·4210358	·464184	2·154315	·9070440	6
24	·4131044	·453620	2·204487	·9106837	36	55	·4212966	·464538	2·152675	·9069215	5
25	·4133693	·453970	2·202784	·9105635	35	56	·4215634	·464891	2·151037	·9067989	4
26	·4136342	·454321	2·201083	·9104432	34	57	·4218272	·465245	2·149402	·9066762	3
27	·4138990	·454672	2·199384	·9103228	33	58	·4220909	·465599	2·147768	·9065535	2
28	·4141638	·455023	2·197687	·9102024	32	59	·4223546	·465953	2·146136	·9064307	1
29	·4144285	·455375	2·195992	·9100819	31	60	·4226183	·466307	2·144506	·9063078	0
30	·4146932	·455726	2·194299	·9099613	30						
′	COSINE.	COTANG.	TANG.	SINE.	′	′	COSINE.	COTANG.	TANG.	SINE.	′

Deg. 65. Deg. 65.

NATURAL SINES AND TANGENTS TO A RADIUS 1.

25 Deg. | 25 Deg.

′	SINE.	TANG.	COTANG.	COSINE.	′	′	SINE.	TANG.	COTANG.	COSINE.	′
0	·4226183	·466307	2·144506	·9063078	60	31	·4307736	·477332	2·094975	·9024600	29
1	·4228819	·466661	2·142879	·9061848	59	32	·4310361	·477689	2·093408	·9023347	28
2	·4231455	·467016	2·141253	·9060618	58	33	·4312986	·478047	2·091843	·9022092	27
3	·4234090	·467370	2·139630	·9059386	57	34	·4315610	·478404	2·090280	·9020838	26
4	·4236725	·467725	2·138008	·9058154	56	35	·4318234	·478762	2·088720	·9019582	25
5	·4239360	·468079	2·136389	·9056922	55	36	·4320857	·479119	2·087161	·9018325	24
6	·4241994	·468434	2·134771	·9055688	54	37	·4323481	·479477	2·085603	·9017068	23
7	·4244628	·468789	2·133155	·9054454	53	38	·4326103	·479835	2·084048	·9015810	22
8	·4247262	·469143	2·131542	·9053219	52	39	·4328726	·480193	2·082495	·9014551	21
9	·4249895	·469498	2·129930	·9051983	51	40	·4331348	·480551	2·080943	·9013292	20
10	·4252528	·469853	2·128321	·9050746	50	41	·4333970	·480909	2·079394	·9012031	19
11	·4255161	·470209	2·126713	·9049509	49	42	·4336591	·481267	2·077846	·9010770	18
12	·4257793	·470564	2·125108	·9048271	48	43	·4339212	·481625	2·076300	·9009508	17
13	·4260425	·470919	2·123504	·9047032	47	44	·4341832	·481984	2·074756	·9008246	16
14	·4263056	·471275	2·121903	·9045792	46	45	·4344453	·482342	2·073214	·9006982	15
15	·4265687	·471630	2·120303	·9044551	45	46	·4347072	·482701	2·071674	·9005718	14
16	·4268318	·471986	2·118705	·9043310	44	47	·4349692	·483060	2·070135	·9004453	13
17	·4270949	·472342	2·117110	·9042068	43	48	·4352311	·483418	2·068599	·9003188	12
18	·4273579	·472697	2·115516	·9040825	42	49	·4354930	·483777	2·067064	·9001921	11
19	·4276208	·473053	2·113924	·9039582	41	50	·4357548	·484136	2·065531	·9000654	10
20	·4278838	·473409	2·112334	·9038338	40	51	·4360166	·484495	2·064000	·8999386	9
21	·4281467	·473765	2·110747	·9037093	39	52	·4362784	·484855	2·062471	·8998117	8
22	·4284095	·474122	2·109161	·9035847	38	53	·4365401	·485214	2·060944	·8996848	7
23	·4286723	·474478	2·107577	·9034600	37	54	·4368018	·485573	2·059418	·8995578	6
24	·4289351	·474834	2·105995	·9033353	36	55	·4370634	·485933	2·057895	·8994307	5
25	·4291979	·475191	2·104415	·9032105	35	56	·4373251	·486293	2·056373	·8993035	4
26	·4294606	·475548	2·102836	·9030856	34	57	·4375866	·486652	2·054853	·8991763	3
27	·4297233	·475904	2·101260	·9029606	33	58	·4378482	·487012	2·053334	·8990489	2
28	·4299859	·476261	2·099686	·9028356	32	59	·4381097	·487372	2·051818	·8989215	1
29	·4302485	·476618	2·098114	·9027105	31	60	·4383711	·487732	2·050303	·8987940	0
30	·4305111	·476975	2·096543	·9025853	30						
′	COSINE.	COTANG.	TANG.	SINE.	′	′	COSINE.	COTANG.	TANG.	SINE.	′

Deg. 64. | Deg. 64.

NATURAL SINES AND TANGENTS TO A RADIUS 1.

26 Deg.

′	SINE.	TANG.	COTANG.	COSINE.	′
0	·4383711	·487732	2·050303	·8987940	60
1	·4386326	·488092	2·048791	·8986665	59
2	·4388940	·488453	2·047280	·8985389	58
3	·4391553	·488813	2·045770	·8984112	57
4	·4394166	·489173	2·044263	·8982834	56
5	·4396779	·489534	2·042757	·8981555	55
6	·4399392	·489894	2·041254	·8980276	54
7	·4402004	·490255	2·039751	·8978996	53
8	·4404615	·490616	2·038251	·8977715	52
9	·4407227	·490977	2·036753	·8976433	51
10	·4409838	·491338	2·035256	·8975151	50
11	·4412448	·491699	2·033761	·8973868	49
12	·4415059	·492061	2·032268	·8972584	48
13	·4417668	·492422	2·030776	·8971299	47
14	·4420278	·492783	2·029287	·8970014	46
15	·4422887	·493145	2·027799	·8968727	45
16	·4425496	·493507	2·026313	·8967440	44
17	·4428104	·493868	2·024828	·8966153	43
18	·4430712	494230	2·023346	·8964864	42
19	·4433319	·494592	2·021865	·8963575	41
20	·4435927	·494954	2·020386	·8962285	40
21	·4438534	·495317	2·018908	·8960994	39
22	·4441140	·495679	2·017433	·8959703	38
23	·4443746	·496041	2·015959	·8958411	37
24	·4446352	·496404	2·014486	·8957118	36
25	·4448957	·496766	2·013016	·8955824	35
26	·4451562	·497129	2·011547	·8954529	34
27	·4454167	·497492	2·010080	·8953234	33
28	·4456771	·497855	2·008615	·8951938	32
29	·4459375	·498218	2·007151	·8950641	31
30	·4461978	·498581	2·005689	·8949344	30
′	COSINE.	COTANG.	TANG.	SINE.	′

Deg. 63.

26 Deg.

′	SINE.	TANG.	COTANG.	COSINE.	′
31	·4464581	·498944	2·004229	·8948045	29
32	·4467184	·499308	2·002771	·8946746	28
33	·4469786	·499671	2·001314	·8945446	27
34	·4472388	·500035	1·999859	·8944146	26
35	·4474990	·500398	1·998405	·8942844	25
36	·4477591	·500762	1·996953	·8941542	24
37	·4480192	·501126	1·995503	·8940240	23
38	·4482792	·501490	1·994055	·8938936	22
39	·4485392	·501854	1·992608	·8937632	21
40	·4487992	·502218	1·991163	·8936326	20
41	·4490591	·502583	1·989720	·8935021	19
42	·4493190	·502947	1·988278	·8933714	18
43	·4495789	·503312	1·986838	·8932406	17
44	·4498387	·503676	1·985400	·8931098	16
45	·4500984	·504041	1·983963	·8929789	15
46	·4503582	·504406	1·982528	·8928480	14
47	·4506179	·504771	1·981095	·8927169	13
48	·4508775	·505136	1·979663	·8925858	12
49	·4511372	·505501	1·978233	·8924546	11
50	·4513967	·505866	1·976805	·8923234	10
51	·4516563	·506232	1·975378	·8921920	9
52	·4519158	·506597	1·973953	·8920606	8
53	·4521753	·506963	1·972529	·8919291	7
54	·4524347	·507329	1·971107	·8917975	6
55	·4526941	·507694	1·969687	·8916659	5
56	·4529535	·508060	1·968268	·8915342	4
57	·4532128	·508426	1·966851	·8914024	3
58	·4534721	·508792	1·965436	·8912705	2
59	·4537313	·509159	1·964022	·8911385	1
60	·4539905	·509525	1·962610	·8910065	0
′	COSINE.	COTANG.	TANG.	SINE.	′

Deg. 63.

NATURAL SINES AND TANGENTS TO A RADIUS 1.

27 Deg. — 27 Deg.

′	SINE.	TANG.	COTANG.	COSINE.	′	′	SINE.	TANG.	COTANG.	COSINE.	′
0	·4539905	·509525	1·962610	·8910065	60	31	·4620066	·520936	1·919618	·8868765	29
1	·4542497	·509891	1·961200	·8908744	59	32	·4622646	·521306	1·918256	·8867420	28
2	·4545088	·510258	1·959791	·8907423	58	33	·4625225	·521676	1·916896	·8866075	27
3	·4547679	·510625	1·958383	·8906100	57	34	·4627804	·522046	1·915537	·8864730	26
4	·4550269	·510991	1·956978	·8904777	56	35	·4630382	·522417	1·914179	·8863383	25
5	·4552859	·511358	1·955573	·8903453	55	36	·4632960	·522787	1·912823	·8862036	24
6	·4555449	·511725	1·954171	·8902128	54	37	·4635538	·523157	1·911469	·8860688	23
7	·4558038	·512093	1·952770	·8900803	53	38	·4638115	·523528	1·910116	·8859339	22
8	·4560627	·512460	1·951371	·8899476	52	39	·4640692	·523899	1·908764	·8857989	21
9	·4563216	·512827	1·949973	·8898149	51	40	·4643269	·524269	1·907414	·8856639	20
10	·4565804	·513195	1·948577	·8896822	50	41	·4645845	·524640	1·906066	·8855288	19
11	·4568392	·513562	1·947182	·8895493	49	42	·4648420	·525011	1·904719	·8853936	18
12	·4570979	·513930	1·945789	·8894164	48	43	·4650996	·525382	1·903373	·8852584	17
13	·4573566	·514298	1·944398	·8892834	47	44	·4653571	·525754	1·902029	·8851230	16
14	·4576153	·514665	1·943008	·8891503	46	45	·4656145	·526125	1·900687	·8849876	15
15	·4578739	·515033	1·941620	·8890171	45	46	·4658719	·526496	1·899346	·8848522	14
16	·4581325	·515401	1·940233	·8888839	44	47	·4661293	·526868	1·898006	·8847166	13
17	·4583910	·515770	1·938848	·8887506	43	48	·4663866	·527240	1·896668	·8845810	12
18	·4586496	·516138	1·937464	·8886172	42	49	·4666439	·527612	1·895332	·8844453	11
19	·4589080	·516506	1·936082	·8884838	41	50	·4669012	·527983	1·893997	·8843095	10
20	·4591665	·516875	1·934702	·8883503	40	51	·4671584	·528356	1·892663	·8841736	9
21	·4594248	·517244	1·933323	·8882166	39	52	·4674156	·528728	1·891331	·8840377	8
22	·4596832	·517612	1·931945	·8880830	38	53	·4676727	·529100	1·890000	·8839017	7
23	·4599415	·517981	1·930569	·8879492	37	54	·4679298	·529472	1·888671	·8837656	6
24	·4601998	·518350	1·929195	·8878154	36	55	·4681869	·529845	1·887343	·8836295	5
25	·4604580	·518719	1·927822	·8876815	35	56	·4684439	·530217	1·886017	·8834933	4
26	·4607162	·519089	1·926451	·8875475	34	57	·4687009	·530590	1·884692	·8833569	3
27	4609744	·519458	1·925081	·8874134	33	58	·4689578	·530963	1·883369	·8832206	2
28	·4612325	·519827	1·923713	·8872793	32	59	·4692147	·531336	1·882047	·8830841	1
29	·4614906	·520197	1·922347	·8871451	31	60	·4694716	·531709	1·880726	·8829476	0
30	·4617486	·520567	1·920982	·8870108	30						
′	COSINE.	COTANG.	TANG.	SINE.	′	′	COSINE.	COTANG.	TANG.	SINE.	′

Deg. 62. — Deg. 62.

NATURAL SINES AND TANGENTS TO A RADIUS 1.

28 Deg.

′	SINE.	TANG.	COTANG.	COSINE.	′
0	·4694716	·531709	1·880726	·8829476	60
1	·4697284	·532082	1·879407	·8828110	59
2	·4699852	·532455	1·878089	·8826743	58
3	·4702419	·532829	1·876773	·8825376	57
4	·4704986	·533202	1·875458	·8824007	56
5	·4707553	·533576	1·874145	·8822638	55
6	·4710119	·533950	1·872833	·8821269	54
7	·4712685	·534324	1·871523	·8819898	53
8	·4715250	·534698	1·870214	·8818527	52
9	·4717815	·535072	1·868906	·8817155	51
10	·4720380	·535446	1·867600	·8815782	50
11	·4722944	·535820	1·866295	·8814409	49
12	·4725508	·536195	1·864992	·8813035	48
13	·4728071	·536569	1·863690	·8811660	47
14	·4730634	·536944	1·862389	·8810284	46
15	·4733197	·537319	1·861090	·8808907	45
16	·4735759	·537694	1·859792	·8807530	44
17	·4738321	·538069	1·858496	·8806152	43
18	·4740882	·538444	1·857201	·8804774	42
19	·4743443	·538819	1·855908	·8803394	41
20	·4746004	·539195	1·854615	·8802014	40
21	·4748564	·539570	1·853325	·8800633	39
22	·4751124	·539946	1·852035	·8799251	38
23	·4753683	·540322	1·850747	·8797869	37
24	·4756242	·540698	1·849461	·8796486	36
25	·4758801	·541074	1·848176	·8795102	35
26	·4761359	·541450	1·846892	·8793717	34
27	·4763917	·541826	1·845609	·8792332	33
28	·4766474	·542202	1·844328	·8790946	32
29	·4769031	·542579	1·843049	·8789559	31
30	·4771588	·542955	1·841770	·8788171	30
′	COSINE.	COTANG.	TANG.	SINE.	′

Deg. 61.

28 Deg.

′	SINE.	TANG.	COTANG.	COSINE.	′
31	·4774144	·543332	1·840494	·8786783	29
32	·4776700	·543709	1·839218	·8785394	28
33	·4779255	·544086	1·837944	·8784004	27
34	·4781810	·544463	1·836671	·8782613	26
35	·4784364	·544840	1·835399	·8781222	25
36	·4786919	·545217	1·834129	·8779830	24
37	·4789472	·545595	1·832861	·8778437	23
38	·4792026	·545972	1·831593	·8777043	22
39	·4794579	·546350	1·830327	·8775649	21
40	·4797131	·546728	1·829062	·8774254	20
41	·4799683	·547106	1·827799	·8772858	19
42	·4802235	·547484	1·826537	·8771462	18
43	·4804786	·547862	1·825276	·8770064	17
44	·4807337	·548240	1·824017	·8768666	16
45	·4809888	·548618	1·822759	·8767268	15
46	·4812438	·548997	1·821502	·8765868	14
47	·4814987	·549375	1·820247	·8764468	13
48	·4817537	·549754	1·818993	·8763067	12
49	·4820086	·550133	1·817740	·8761665	11
50	·4822634	·550512	1·816489	·8760263	10
51	·4825182	·550891	1·815239	·8758859	9
52	·4827730	·551270	1·813990	·8757455	8
53	·4830277	·551650	1·812743	·8756051	7
54	·4832824	·552029	1·811496	·8754645	6
55	·4835370	·552409	1·810252	·8753239	5
56	·4837916	·552789	1·809008	·8751832	4
57	·4840462	·553168	1·807766	·8750425	3
58	·4843007	·553548	1·806525	·8749016	2
59	·4845552	·553928	1·805286	·8747607	1
60	·4848096	·554309	1·804047	·8746197	0
′	COSINE.	COTANG.	TANG.	SINE.	′

Deg. 61.

NATURAL SINES AND TANGENTS TO A RADIUS 1.

29 Deg.						29 Deg.					
′	SINE.	TANG.	COTANG.	COSINE.	′	′	SINE.	TANG.	COTANG.	COSINE.	′
0	·4848096	·554309	1·804047	·8746197	60	31	·4926767	·566156	1·766295	·8702124	29
1	·4850640	·554689	1·802810	·8744786	59	32	·4929298	·566541	1·765097	·8700691	28
2	·4853184	·555069	1·801575	·8743375	58	33	·4931829	·566925	1·763900	·8699256	27
3	·4855727	·555450	1·800340	·8741963	57	34	·4934359	·567309	1·762705	·8697821	26
4	·4858270	·555831	1·799107	·8740550	56	35	·4936889	·567694	1·761511	·8696386	25
5	·4860812	·556211	1·797875	·8739137	55	36	·4939419	·568079	1·760318	·8694949	24
6	·4863354	·556592	1·796645	·8737722	54	37	·4941948	·568463	1·759126	·8693512	23
7	·4865895	·556973	1·795416	·8736307	53	38	·4944476	·568848	1·757936	·8692074	22
8	·4868436	·557355	1·794188	·8734891	52	39	·4947005	·569233	1·756747	·8690636	21
9	·4870977	·557736	1·792961	·8733475	51	40	·4949532	·569619	1·755559	·8689196	20
10	·4873517	·558117	1·791736	·8732058	50	41	·4952060	·570004	1·754372	·8687756	19
11	·4876057	·558499	1·790512	·8730640	49	42	·4954587	·570389	1·753186	·8686315	18
12	·4878597	·558881	1·789289	·8729221	48	43	·4957113	·570775	1·752002	·8684874	17
13	·4881136	·559262	1·788067	·8727801	47	44	·4959639	·571161	1·750819	·8683431	16
14	·4883674	·559644	1·786847	·8726381	46	45	·4962165	·571547	1·749637	·8681988	15
15	·4886212	·560026	1·785628	·8724960	45	46	·4964690	·571933	1·748456	·8680544	14
16	·4888750	·560409	1·784410	·8723538	44	47	·4967215	·572319	1·747276	·8679100	13
17	4891288	·560791	1·783194	·8722116	43	48	·4969740	·572705	1·746098	·8677655	12
18	·4893825	·561173	1·781979	·8720693	42	49	·4972264	·573091	1·744921	·8676209	11
19	·4896361	·561556	1·780765	·8719269	41	50	·4974787	·573478	1·743745	·8674762	10
20	·4898897	·561939	1·779552	·8717844	40	51	·4977310	·573864	1·742570	·8673314	9
21	·4901433	·562321	1·778340	·8716419	39	52	·4979833	·574251	1·741396	·8671866	8
22	·4903968	·562704	1·777130	·8714993	38	53	·4982355	·574638	1·740224	·8670417	7
23	·4906503	·563087	1·775921	·8713566	37	54	·4984877	·575025	1·739053	·8668967	6
24	·4909038	·563471	1·774714	·8712138	36	55	·4987399	·575412	1·737883	·8667517	5
25	·4911572	·563854	1·773507	·8710710	35	56	·4989920	·575799	1·736714	·8666066	4
26	·4914105	·564237	1·772302	·8709281	34	57	·4992441	·576187	1·735546	·8664614	3
27	·4916638	·564621	1·771098	·8707851	33	58	·4994961	·576574	1·734380	·8663161	2
28	·4919171	·565005	1·769895	·8706420	32	59	·4997481	·576962	1·733214	·8661708	1
29	·4921704	·565388	1·768694	·8704989	31	60	·5000000	·577350	1·732050	·8660254	0
30	·4924236	·565772	1·767494	·8703557	30						
′	COSINE.	COTANG.	TANG.	SINE.	′	′	COSINE.	COTANG.	TANG.	SINE.	′
				Deg. 60.						Deg. 60.	

NATURAL SINES AND TANGENTS TO A RADIUS 1.

30 Deg. | 30 Deg.

′	SINE.	TANG.	COTANG.	COSINE.	′	′	SINE.	TANG.	COTANG.	COSINE.	′
0	·5000000	·577350	1·732050	·8660254	60	31	·5077890	·589436	1·696534	·8614815	29
1	·5002519	577738	1·730887	·8658799	59	32	·5080396	·589828	1·695406	·8613337	28
2	·5005037	·578126	1·729726	·8657344	58	33	5082901	·590221	1·694280	·8611859	27
3	·5007556	·578514	1·728565	·8655887	57	34	·5085406	·590613	1·693155	·8610380	26
4	·5010073	·578902	1·727406	·8654430	56	35	·5087910	·591005	1·692030	·8608901	25
5	·5012591	·579291	1·726247	·8652973	55	36	·5090414	·591398	1·690907	·8607420	24
6	·5015107	·579679	1·725090	8651514	54	37	·5092918	·591791	1·689785	·8605939	23
7	·5017624	·580068	1·723934	8650055	53	38	·5095421	·592183	1·688664	·8604457	22
8	·5020140	·580457	1·722779	8648595	52	39	·5097924	·592576	1·687544	·8602975	21
9	·5022655	·580846	1·721626	8647134	51	40	·5100426	·592969	1·686426	·8601491	20
10	·5025170	·581235	1·720473	8645673	50	41	·5102928	·593363	1·685308	·8600007	19
11	·5027685	·581624	1·719322	8644211	49	42	·5105429	·593756	1·684191	·8598523	18
12	·5030199	·582013	1·718172	8642748	48	43	·5107930	·594150	1·683076	·8597037	17
13	·5032713	·582403	1·717023	8641284	47	44	·5110431	·594543	1·681962	·8595551	16
14	·5035227	·582793	1·715875	8639820	46	45	·5112931	·594937	1·680848	·8594064	15
15	·5037740	·583182	1·714728	8638355	45	46	·5115431	·595331	1·679736	·8592576	14
16	·5040252	·583572	1·713582	8636889	44	47	·5117930	·595725	1·678625	·8591088	13
17	·5042765	·583962	1·712438	8635423	43	48	·5120429	·596119	1·677515	·8589599	12
18	·5045276	·584352	1·711294	8633956	42	49	·5122927	·596514	1·6764[illegible]6	·8588109	11
19	·5047788	·584743	1·710152	8632488	41	50	·5125425	596908	1·6752[illegible]8	·8586619	10
20	·5050298	·585133	1·709011	8631019	40	51	5127923	·597303	1·6741[illegible]2	·8585127	9
21	·5052809	·585524	1·707871	8629549	39	52	5130420	·597697	1·6730[illegible]	·8583635	8
22	·5055319	·585914	1·706732	8628079	38	53	5132916	·598092	1·6719[illegible]1	·8582143	7
23	·5057828	·586305	1·705595	8626608	37	54	·5135413	598487	1·6708[illegible]8	·8580649	6
24	·5060338	·586696	1·704458	8625137	36	55	·5137908	598882	1·6697[illegible]5	·8579155	5
25	·5062846	·587087	1·703323	8623664	35	56	·5140404	·599278	1·668674	·8577660	4
26	·5065355	·587478	1·702189	8622191	34	57	·5142899	599673	1·6675[illegible]4	·8576164	3
27	·5067863	·587870	1·701055	8620717	33	58	·5145393	600069	1·6664[illegible]4	·8574668	2
28	·5070370	·588261	1·699923	8619243	32	59	·5147887	600464	1·665376	·8573171	1
29	·5072877	·588653	1·698792	8617768	31	60	·5150381	600860	1·664279	·8571673	0
30	·5075384	·589045	1·697663	8616292	30						
′	COSINE.	COTANG.	TANG.	SINE.	′	′	COSINE.	COTANG.	TANG.	SINE.	′

Deg. 59. | Deg. 59.

NATURAL SINES AND TANGENTS TO A RADIUS 1.

31 Deg.

′	SINE.	TANG.	COTANG.	COSINE.	′
0	·5150381	·600860	1·664279	·8571673	60
1	·5152874	·601256	1·663183	·8570174	59
2	·5155367	·601652	1·662088	8568675	58
3	·5157859	·602049	1·660994	8567175	57
4	·5160351	·602445	1·659901	8565674	56
5	·5162842	·602841	1·658809	8564173	55
6	·5165333	·603238	1·657718	8562671	54
7	·5167824	·603635	1·656629	8561168	53
8	5170314	·604032	1·655540	8559664	52
9	5172804	·604429	1·654452	8558160	51
10	·5175293	·604826	1·653366	8556655	50
11	·5177782	·605224	1·652280	8555149	49
12	·5180270	·605621	1·651196	8553643	48
13	·5182758	·606019	1·650112	8552135	47
14	5185246	·606417	1·649030	8550627	46
15	5187733	·606814	1·647949	8549119	45
16	5190219	607213	1·646868	8547609	44
17	5192705	·607611	1·645789	8546099	43
18	5195191	608009	1·644711	8544588	42
19	5197676	·608408	1·643633	8543077	41
20	5200161	608806	1·642557	8541564	40
21	5202646	609205	1·641482	8540051	39
22	5205130	609604	1·640408	8538538	38
23	5207613	610003	1·639335	8537023	37
24	5210096	610402	1·638263	8535508	36
25	5212579	610801	1·637191	8533992	35
26	5215061	611201	1·636121	8532475	34
27	5217543	611601	1·635052	8530958	33
28	5220024	612000	1·633984	8529440	32
29	5222505	612400	1·632917	8527921	31
30	5224986	612800	1·631851	8526402	30
′	COSINE.	COTANG.	TANG.	SINE.	′

Deg. 58.

31 Deg.

′	SINE.	TANG.	COTANG.	COSINE.	′
31	·5227466	·613201	1·630786	·8524881	29
32	·5229945	·613601	1·629722	·8523360	28
33	·5232424	·614001	1·628659	·8521839	27
34	·5234903	·614402	1·627597	·8520316	26
35	·5237381	·614803	1·626536	·8518793	25
36	·5239859	·615204	1·625476	·8517269	24
37	·5242336	·615605	1·624417	·8515745	23
38	·5244813	·616006	1·623359	·8514219	22
39	·5247290	·616407	1·622302	·8512693	21
40	·5249766	616809	1·621246	·8511167	20
41	·5252241	·617210	1·620192	·8509639	19
42	·5254717	·617612	1·619138	·8508111	18
43	·5257191	·618014	1·618085	·8506582	17
44	·5259665	·618416	1·617033	·8505053	16
45	·5262139	·618818	1·615982	·8503522	15
46	·5264613	·619221	1·614932	·8501991	14
47	·5267085	·619623	1·613882	·8500459	13
48	·5269558	·620026	1·612834	·8498927	12
49	·5272030	·620429	1·611787	·8497394	11
50	·5274502	·620832	1·610741	·8495860	10
51	·5276973	·621235	1·609696	·8494325	9
52	·5279443	·621638	1·608652	·8492790	8
53	·5281914	·622041	1·607609	·8491254	7
54	·5284383	·622445	1·606567	·8489717	6
55	·5286853	·622848	1·605526	·8488179	5
56	·5289322	·623252	1·604485	·8486641	4
57	·5291790	·623656	1·603446	·8485102	3
58	·5294258	·624060	1·602408	·8483562	2
59	·5296726	·624465	1·601370	·8482022	1
60	·5299193	·624869	1·600334	·8480481	0
′	COSINE.	COTANG.	TANG.	SINE.	′

Deg. 58

NATURAL SINES AND TANGENTS TO A RADIUS 1.

32 DEG. 32 DEG.

′	SINE.	TANG.	COTANG.	COSINE.	′.	′	SINE.	TANG.	COTANG.	COSINE.	′
0	·5299193	·624869	1·600334	·8480481	60	31	·5375449	·637479	1·568678	·8432351	29
1	·5301659	·625273	1·599299	·8478939	59	32	·5377902	·637888	1·567672	·8430787	28
2	·5304125	·625678	1·598264	·8477397	58	33	·5380354	·638297	1·566666	·8429222	27
3	·5306591	·626083	1·597231	·8475853	57	34	·5382806	·638707	1·565662	·8427657	26
4	·5309057	·626488	1·596198	·8474309	56	35	·5385257	·639116	1·564659	·8426091	25
5	·5311521	·626893	1·595167	·8472765	55	36	·5387708	·639526	1·563656	·8424524	24
6	·5313986	·627298	1·594136	·8471219	54	37	·5390158	·639936	1·562654	·8422956	23
7	·5316450	·627704	1·593107	·8469673	53	38	·5392608	·640346	1·561654	·8421388	22
8	·5318913	·628109	1·592078	·8468126	52	39	·5395058	·640756	1·560654	·8419819	21
9	·5321376	·628515	1·591050	·8466579	51	40	·5397507	·641167	1·559655	·8418249	20
10	·5323839	·628921	1·590023	·8465030	50	41	·5399955	·641577	1·558657	·8416679	19
11	·5326301	·629327	1·588997	·8463481	49	42	·5402403	·641988	1·557660	·8415108	18
12	·5328763	·629733	1·587973	·8461932	48	43	·5404851	·642399	1·556663	·8413536	17
13	·5331224	·630139	1·586949	·8460381	47	44	·5407298	·642810	1·555668	·8411963	16
14	·5333685	·630546	1·585926	·8458830	46	45	·5409745	·643221	1·554674	·8410390	15
15	·5336145	·630953	1·584904	·8457278	45	46	·5412191	·643632	1·553680	·8408816	14
16	·5338605	·631359	1·583883	·8455726	44	47	·5414637	·644044	1·552688	·8407241	13
17	·5341065	·631766	1·582862	·8454172	43	48	·5417082	·644456	1·551696	·8405666	12
18	·5343523	·632173	1·581843	·8452618	42	49	·5419527	·644867	1·550705	·8404090	11
19	·5345982	·632581	1·580825	·8451064	41	50	·5421971	·645279	1·549715	·8402513	10
20	·5348440	·632988	1·579807	·8449508	40	51	·5424415	·645691	1·548726	·8400936	9
21	·5350898	·633395	1·578791	·8447952	39	52	·5426859	·646104	1·547738	·8399357	8
22	·5353355	·633803	1·577776	·8446395	38	53	·5429302	·646516	1·546751	·8397778	7
23	·5355812	·634211	1·576761	·8444838	37	54	·5431744	·646929	1·545764	·8396199	6
24	·5358268	·634619	1·575747	·8443279	36	55	·5434187	·647341	1·544779	·8394618	5
25	·5360724	·635027	1·574735	·8441720	35	56	·5436628	·647754	1·543794	·8393037	4
26	·5363179	·635435	1·573723	·8440161	34	57	·5439069	·648167	1·542810	·8391455	3
27	·5365634	·635844	1·572712	·8438600	33	58	·5441510	·648580	1·541828	·8389873	2
28	·5368089	·636252	1·571702	·8437039	32	59	·5443951	·648994	1·540846	·8388290	1
29	·5370543	·636661	1·570693	·8435477	31	60	·5446390	·649407	1·539865	·8386706	0
30	·5372996	·637070	1·569685	·8433914	30						
′	COSINE.	COTANG.	TANG.	SINE.	′	′	COSINE.	COTANG.	TANG.	SINE.	′

DEG. 57. DEG. 57

NATURAL SINES AND TANGENTS TO A RADIUS 1.

33 Deg. | 33 Deg.

′	SINE.	TANG.	COTANG.	COSINE.	′	′	SINE.	TANG.	COTANG.	COSINE.	′
0	·5446390	·649407	1·539865	·8386706	60	31	5521795	·662304	1·509880	·8337252	29
1	·5448830	·649821	1·538884	·8385121	59	32	5624220	·662722	1·508927	·8335646	28
2	·5451269	·650235	1·537905	·8383536	58	33	·5526645	·663141	1·507974	·8334038	27
3	·5453707	·650649	1·536927	·8381950	57	34	·5529069	·663560	1·507022	·8332430	26
4	·5456145	·651063	1·535949	·8380363	56	35	·5531492	·663979	1·506071	·8330822	25
5	·5458583	·651477	1·534972	·8378775	55	36	·5533915	·664398	1·505121	·8329212	24
6	·5461020	·651891	1·533996	·8377187	54	37	·5536338	·664817	1·504171	·8327602	23
7	·5463456	·652306	1·533021	·8375598	53	38	·5538760	·665237	1·503222	·8325991	22
8	·5465892	·652721	1·532047	·8374009	52	39	·5541182	·665657	1·502275	·8324380	21
9	·5468328	·653136	1·531074	·8372418	51	40	·5543603	·666076	1·501328	·8322768	20
10	·5470763	·653551	1·530102	·8370827	50	41	·5546024	·666496	1·500382	·8321155	19
11	·5473198	·653966	1·529130	·8369236	49	42	·5548444	·666917	1·499436	·8319541	18
12	·5475632	·654381	1·528160	·8367643	48	43	·5550864	·667337	1·498492	·8317927	17
13	·5478066	·654797	1·527190	·8366050	47	44	·5553283	·667758	1·497548	8316312	16
14	·5480499	·655212	1·526221	·8364456	46	45	·5555702	·668178	1·496605	·8314696	15
15	·5482932	·655628	1·525253	·8362862	45	46	·5558121	·668599	1·495663	·8313080	14
16	·5485365	·656044	1·524286	·8361266	44	47	·5560539	·669020	1·494722	·8311463	13
17	·5487797	·656460	1·523320	·8359670	43	48	·5562956	·669441	1·493782	·8309845	12
18	·5490228	·656877	1·522354	·8358074	42	49	·5565373	·669863	1·492842	·8308226	11
19	·5492659	·657293	1·521389	·8356476	41	50	·5567790	·670284	1·491903	·8306607	10
20	·5495090	·657710	1·520426	·8354878	40	51	·5570206	·670706	1·490965	·8304987	9
21	·5497520	·658127	1·519463	·8353279	39	52	·5572621	·671128	1·490028	·8303366	8
22	·5499950	·658544	1·518501	·8351680	38	53	·5575036	·671550	1·489092	8301745	7
23	·5502379	·658961	1·517540	·8350080	37	54	·5577451	·671972	1·488157	8300123	6
24	·5504807	·659378	1·516579	·8348479	36	55	·5579865	·672394	1·487222	8298500	5
25	·5507236	·659796	1·515620	·8346877	35	56	·5582279	·672816	1·486288	·8296877	4
26	·5509663	·660213	1·514661	·8345275	34	57	·5584692	·673239	1·485355	·8295252	3
27	·5512091	·660631	1·513703	·8343672	33	58	·5587105	·673662	1·484423	·8293628	2
28	·5514518	·661049	1·512746	·8342068	32	59	·5589517	·674085	1·483491	·8292002	1
29	·5516944	·661467	1·511790	8340463	31	60	·5591929	·674508	1·482561	·8290376	0
30	·5519370	·661885	1·510835	·8338858	30						
′	COSINE.	COTANG.	TANG.	SINE.	′	′	COSINE.	COTANG.	TANG.	SINE.	′

Deg. 56. | Deg. 56.

NATURAL SINES AND TANGENTS TO A RADIUS 1.

34 Deg.

′	SINE.	TANG.	COTANG.	COSINE.	′
0	·5591929	·674508	1·482561	·8290376	60
1	·5594340	·674931	1·481631	·8288749	59
2	·5596751	·675355	1·480702	·8287121	58
3	·5599162	·675779	1·479773	·8285493	57
4	·5601572	·676202	1·478846	·8283864	56
5	·5603981	·676626	1·477919	·8282234	55
6	·5606390	·677050	1·476993	·8280603	54
7	·5608798	·677475	1·476068	·8278972	53
8	·5611206	·677899	1·475144	·8277340	52
9	·5613614	·678324	1·474221	·8275708	51
10	·5616021	·678749	1·473298	·8274074	50
11	·5618428	·679174	1·472376	·8272440	49
12	·5620834	·679599	1·471455	·8270806	48
13	·5623239	·680024	1·470535	·8269170	47
14	·5625645	·680450	1·469615	·8267534	46
15	·5628049	·680875	1·468696	·8265897	45
16	·5630453	·681301	1·467778	·8264260	44
17	·5632857	·681727	1·466861	·8262622	43
18	·5635260	·682153	1·465945	·8260983	42
19	·5637663	·682580	1·465029	·8259343	41
20	·5640066	·683006	1·464114	·8257703	40
21	·5642467	·683433	1·463200	·8256062	39
22	·5644869	·683860	1·462287	·8254420	38
23	·5647270	·684287	1·461374	·8252778	37
24	·5649670	·684714	1·460463	·8251135	36
25	·5652070	·685141	1·459552	·8249491	35
26	·5654469	·685569	1·458642	·8247847	34
27	·5656868	·685996	1·457732	·8246202	33
28	·5659267	·686424	1·456824	·8244556	32
29	·5661665	·686852	1·455916	·8242909	31
30	·5664062	·687281	1·455009	·8241262	30
′	COSINE.	COTANG.	TANG.	SINE.	′

Deg. 55

34 Deg.

′	SINE.	TANG.	COTANG.	COSINE.	′
31	·5666459	·687709	1·454102	·8239614	29
32	·5668856	·688137	1·453197	·8237965	28
33	·5671252	·688566	1·452292	·8236316	27
34	·5673648	·688995	1·451388	·8234666	26
35	·5676043	·689424	1·450485	·8233015	25
36	·5678437	·689853	1·449582	·8231364	24
37	·5680832	·690283	1·448680	·8229712	23
38	·5683225	·690712	1·447779	·8228059	22
39	·5685619	·691142	1·446879	·8226405	21
40	·5688011	·691572	1·445980	·8224751	20
41	·5690403	·692002	1·445081	·8223096	19
42	·5692795	·692432	1·444183	·8221440	18
43	·5695187	·692863	1·443286	·8219784	17
44	·5697577	·693293	1·442389	·8218127	16
45	·5699968	·693724	1·441494	·8216469	15
46	·5702357	·694155	1·440599	·8214811	14
47	·5704747	·694586	1·439704	·8213152	13
48	·5707136	·695018	1·438811	·8211492	12
49	·5709524	·695449	1·437918	·8209832	11
50	·5711912	·695881	1·437026	·8208170	10
51	·5714299	·696313	1·436135	·8206509	9
52	·5716686	·696745	1·435245	·8204846	8
53	·5719073	·697177	1·434355	·8203183	7
54	·5721459	·697609	1·433466	·8201519	6
55	·5723844	·698042	1·432578	·8199854	5
56	·5726229	·698474	1·431690	·8198189	4
57	·5728614	·698907	1·430803	·8196523	3
58	·5730998	·699340	1·429917	·8194856	2
59	·5733381	·699774	1·429032	·8193189	1
60	·5735764	·700207	1·428148	·8191520	0
′	COSINE.	COTANG.	TANG.	SINE.	′

Deg. 55.

NATURAL SINES AND TANGENTS TO A RADIUS 1.

35 Deg. | 35 Deg.

′	SINE.	TANG.	COTANG.	COSINE.	′	′	SINE.	TANG.	COTANG.	COSINE.	′
0	·5735764	·700207	1·428148	·8191520	60	31	·5809397	·713732	1·401086	·8139466	29
1	·5738147	·700641	1·427264	·8189852	59	32	·5811765	·714171	1·400224	·8137775	28
2	·5740529	·701074	1·426381	·8188182	58	33	·5814132	·714610	1·399363	·8136084	27
3	·5742911	·701508	1·425498	·8186512	57	34	·5816498	·715050	1·398503	·8134393	26
4	·5745292	·701943	1·424617	·8184841	56	35	·5818864	·715489	1·397644	·8132701	25
5	·5747672	·702377	1·423736	·8183169	55	36	·5821230	·715929	1·396785	·8131008	24
6	·5750053	·702811	1·422856	·8181497	54	37	·5823595	·716369	1·395927	·8129314	23
7	·5752432	·703246	1·421976	·8179824	53	38	·5825959	·716810	1·395069	·8127620	22
8	·5754811	·703681	1·421097	·8178151	52	39	·5828323	·717250	1·394213	·8125925	21
9	·5757190	·704116	1·420220	·8176476	51	40	·5830687	·717691	1·393357	·8124229	20
10	·5759568	·704551	1·419342	·8174801	50	41	·5833050	·718131	1·392501	·8122532	19
11	·5761946	·704986	1·418466	·8173125	49	42	·5835412	·718572	1·391647	·8120835	18
12	·5764323	·705422	1·417590	·8171449	48	43	·5837774	·719014	1·390793	·8119137	17
13	·5766700	·705858	1·416715	·8169772	47	44	·5840136	·719455	1·389940	·8117439	16
14	·5769076	·706294	1·415840	·8168094	46	45	·5842497	·719897	1·389087	·8115740	15
15	·5771452	·706730	1·414967	·8166416	45	46	·5844857	·720338	1·388235	·8114040	14
16	·5773827	·707166	1·414094	·8164736	44	47	·5847217	·720780	1·387384	·8112339	13
17	·5776202	·707602	1·413222	·8163056	43	48	·5849577	·721222	1·386534	·8110638	12
18	·5778576	·708039	1·412350	·8161376	42	49	·5851936	·721665	1·385684	·8108936	11
19	·5780950	·708476	1·411479	·8159695	41	50	·5854294	·722107	1·384835	·8107234	10
20	·5783323	·708913	1·410609	·8158013	40	51	·5856652	·722550	1·383986	·8105530	9
21	·5785696	·709350	1·409740	·8156330	39	52	·5859010	·722993	1·383139	·8103826	8
22	·5788069	·709787	1·408871	·8154647	38	53	·5861367	·723436	1·382292	·8102122	7
23	·5790440	·710225	1·408003	·8152963	37	54	·5863724	·723879	1·381445	·8100416	6
24	·5792812	·710663	1·407136	·8151278	36	55	·5866080	·724322	1·380600	·8098710	5
25	·5795183	·711100	1·406270	·8149593	35	56	·5868435	·724766	1·379755	·8097004	4
26	·5797553	·711539	1·405404	·8147906	34	57	·5870790	·725210	1·378910	·8095296	3
27	·5799923	·711977	1·404539	·8146220	33	58	·5873145	·725654	1·378067	·8093588	2
28	·5802292	·712415	1·403674	·8144532	32	59	·5875499	·726098	1·377224	·8091879	1
29	·5804661	·712854	1·402811	·8142844	31	60	·5877853	·726542	1·376381	·8090170	0
30	·5807030	·713293	1·401948	·8141155	30						
′	COSINE.	COTANG.	TANG.	SINE.	′	′	COSINE.	COTANG.	TANG.	SINE.	′

Deg. 54. | Deg. 54.

NATURAL SINES AND TANGENTS TO A RADIUS 1.

36 Deg. | 36 Deg.

′	SINE.	TANG.	COTANG.	COSINE.	′	′	SINE.	TANG.	COTANG.	COSINE.	′
0	·5877853	·726542	1·376381	·8090170	60	31	·5950566	·740411	1·350600	·8036838	29
1	·5880206	·726987	1·375540	·8088460	59	32	·5952904	·740861	1·349779	·8035107	28
2	·5882558	·727431	1·374699	·8086749	58	33	·5955241	·741312	1·348958	·8033375	27
3	·5884910	·727876	1·373859	·8085037	57	34	·5957577	·741763	1·348139	·8031642	26
4	·5887262	·728321	1·373019	·8083325	56	35	·5959913	·742214	1·347319	·8029909	25
5	·5889613	·728767	1·372180	·8081612	55	36	·5962249	·742665	1·346501	·8028175	24
6	·5891964	·729212	1·371342	·8079899	54	37	·5964584	·743117	1·345683	·8026440	23
7	·5894314	·729658	1·370504	·8078185	53	38	·5966918	·743568	1·344865	·8024705	22
8	·5896663	·730104	1·369667	·8076470	52	39	·5969252	·744020	1·344049	·8022969	21
9	·5899012	·730550	1·368831	·8074754	51	40	·5971586	·744472	1·343233	·8021232	20
10	·5901361	·730996	1·367995	·8073038	50	41	·5973919	·744924	1·342417	·8019495	19
11	·5903709	·731442	1·367161	·8071321	49	42	·5976251	·745377	1·341602	·8017756	18
12	·5906057	·731889	1·366326	·8069603	48	43	·5978583	·745829	1·340788	·8016018	17
13	·5908404	·732336	1·365493	·8067885	47	44	·5980915	·746282	1·339975	·8014278	16
14	·5910750	·732783	1·364660	·8066166	46	45	·5983246	·746735	1·339162	·8012538	15
15	·5913096	·733230	1·363827	·8064446	45	46	·5985577	·747188	1·338350	·8010797	14
16	·5915442	·733677	1·362996	·8062726	44	47	·5987906	·747642	1·337538	·8009056	13
17	·5917787	·734125	1·362165	·8061005	43	48	·5990236	·748095	1·336727	·8007314	12
18	·5920132	·734573	1·361335	·8059283	42	49	·5992565	·748549	1·335917	·8005571	11
19	·5922476	·735021	1·360505	·8057560	41	50	·5994893	·749003	1·335107	·8003827	10
20	·5924819	·735469	1·359676	·8055837	40	51	·5997221	·749457	1·334298	·8002083	9
21	·5927163	·735917	1·358848	·8054113	39	52	·5999549	·749911	1·333490	·8000338	8
22	·5929505	·736366	1·358020	·8052389	38	53	·6001876	·750366	1·332682	·7998593	7
23	·5931847	·736814	1·357193	·8050664	37	54	·6004202	·750821	1·331875	·7996847	6
24	·5934189	·737263	1·356367	·8048938	36	55	·6006528	·751276	1·331068	·7995100	5
25	·5936530	·737712	1·355541	·8047211	35	56	·6008854	·751731	1·330262	·7993352	4
26	5938871	·738162	1·354716	·8045484	34	57	·6011179	·752186	1·329457	·7991604	3
27	·5941211	·738611	1·353891	·8043756	33	58	·6013503	·752642	1·328653	·7989855	2
28	·5943550	·739061	1·353068	·8042028	32	59	·6015827	·753098	1·327848	·7988105	1
29	·5945889	·739511	1·352244	·8040299	31	60	·6018150	·753554	1·327044	·7986355	0
30	·5948228	·739961	1·351422	·8038569	30						
′	COSINE.	COTANG.	TANG.	SINE.	′	′	COSINE.	COTANG.	TANG.	SINE.	′

Deg. 53. | Deg. 53.

NATURAL SINES AND TANGENTS TO A RADIUS 1.

37 Deg.

′	SINE.	TANG.	COTANG.	COSINE.	′	′	SINE.	TANG.	COTANG.	COSINE.	′
0	·6018150	·753554	1·327044	·7986355	60	31	·6089922	·767789	1·302440	·7931762	29
1	·6020473	·754010	1·326242	·7984604	59	32	·6092229	·768251	1·301656	·7929990	28
2	·6022795	·754466	1·325439	·7982853	58	33	·6094535	·768714	1·300873	·7928218	27
3	·6025117	·754923	1·324638	·7981100	57	34	·6096841	·769177	1·300090	·7926445	26
4	·6027439	·755379	1·323837	·7979347	56	35	·6099147	·769640	1·299308	·7924671	25
5	·6029760	·755836	1·323036	·7977594	55	36	·6101452	·770103	1·298526	·7922896	24
6	·6032080	·756294	1·322237	·7975839	54	37	·6103756	·770567	1·297745	·7921121	23
7	·6034400	·756751	1·321437	·7974084	53	38	·6106060	·771030	1·296964	·7919345	22
8	·6036719	·757209	1·320639	·7972329	52	39	·6108363	·771494	1·296185	·7917569	21
9	·6039038	·757666	1·319841	·7970572	51	40	·6110666	·771958	1·295405	·7915792	20
10	·6041356	·758124	1·319044	·7968815	50	41	·6112969	·772423	1·294627	·7914014	19
11	·6043674	·758582	1·318247	·7967058	49	42	·6115270	·772887	1·293848	·7912235	18
12	·6045991	·759041	1·317451	·7965299	48	43	·6117572	·773352	1·293071	·7910456	17
13	·6048308	·759499	1·316655	·7963540	47	44	·6119873	·773817	1·292293	·7908676	16
14	·6050624	·759958	1·315861	·7961780	46	45	·6122173	·774282	1·291517	·7906896	15
15	·6052940	·760417	1·315066	·7960020	45	46	·6124473	·774748	1·290742	·7905115	14
16	·6055255	·760876	1·314273	·7958259	44	47	·6126772	·775213	1·289966	·7903333	13
17	·6057570	·761336	1·313480	·7956497	43	48	·6129071	·775679	1·289192	·7901550	12
18	·6059884	·761795	1·312687	·7954735	42	49	·6131369	·776145	1·288418	·7899767	11
19	·6062198	·762255	1·311895	·7952972	41	50	·6133666	·776611	1·287644	·7897983	10
20	·6064511	·762715	1·311104	·7951208	40	51	·6135964	·777078	1·286871	·7896198	9
21	·6066824	·763175	1·310314	·7949444	39	52	·6138260	·777544	1·286099	·7894413	8
22	·6069136	·763636	1·309523	·7947678	38	53	·6140556	·778011	1·285327	·7892627	7
23	·6071447	·764097	1·308734	·7945913	37	54	·6142852	·778478	1·284556	·7890841	6
24	·6073758	·764557	1·307945	·7944146	36	55	·6145147	·778946	1·283786	·7889054	5
25	·6076069	·765018	1·307157	·7942379	35	56	·6147442	·779413	1·283016	·7887266	4
26	·6078379	·765480	1·306369	·7940611	34	57	·6149736	·779881	1·282246	·7885477	3
27	·6080689	·765941	1·305582	·7938843	33	58	·6152029	·780349	1·281477	·7883688	2
28	·6082998	·766403	1·304796	·7937074	32	59	·6154322	·780817	1·280709	·7881898	1
29	·6085306	·766864	1·304010	·7935304	31	60	·6156615	·781285	1·279941	·7880108	0
30	·6087614	·767327	1·303225	·7933533	30						
′	COSINE.	COTANG.	TANG.	SINE.	′	′	COSINE.	COTANG.	TANG.	SINE.	′

Deg. 52.

NATURAL SINES AND TANGENTS TO A RADIUS 1.

38 Deg.

′	SINE.	TANG.	COTANG.	COSINE.	′
0	·6156615	·781285	1·279941	·7880108	60
1	·6158907	·781754	1·279174	·7878316	59
2	·6161198	·782222	1·278407	·7876524	58
3	·6163489	·782691	1·277641	·7874732	57
4	·6165780	·783161	1·276876	·7872939	56
5	·6168069	·783630	1·276111	·7871145	55
6	·6170359	·784100	1·275347	·7869350	54
7	·6172648	·784570	1·274583	·7867555	53
8	·6174936	·785040	1·273820	·7865759	52
9	·6177224	·785510	1·273057	·7863963	51
10	·6179511	·785980	1·272295	·7862165	50
11	·6181798	·786451	1·271534	·7860367	49
12	·6184084	·786922	1·270773	·7858569	48
13	·6186370	·787393	1·270013	·7856770	47
14	·6188655	·787864	1·269253	·7854970	46
15	·6190939	·788336	1·268494	·7853169	45
16	·6193224	·788808	1·267735	·7851368	44
17	·6195507	·789280	1·266977	·7849566	43
18	·6197790	·789752	1·266219	·7847764	42
19	·6200073	·790224	1·265462	·7845961	41
20	·6202355	·790697	1·264706	·7844157	40
21	·6204636	·791170	1·263950	·7842352	39
22	·6206917	·791643	1·263195	·7840547	38
23	·6209198	·792116	1·262440	·7838741	37
24	·6211478	·792590	1·261686	·7836935	36
25	·6213757	·793064	1·260932	·7835127	35
26	·6216036	·793537	1·260179	·7833320	34
27	·6218314	·794012	1·259426	·7831511	33
28	·6220592	·794486	1·258674	·7829702	32
29	·6222870	·794961	1·257923	·7827892	31
30	·6225146	·795435	1·257172	·7826082	30
′	COSINE.	COTANG.	TANG.	SINE.	′

Deg. 51.

38 Deg.

′	SINE.	TANG.	COTANG.	COSINE.	′
31	·6227423	·795911	1·256421	·7824270	29
32	·6229698	·796386	1·255672	·7822459	28
33	·6231974	·796861	1·254922	·7820646	27
34	·6234248	·797337	1·254174	·7818833	26
35	·6236522	·797813	1·253426	·7817019	25
36	·6238796	·798289	1·252678	·7815205	24
37	·6241069	·798765	1·251931	·7813390	23
38	·6243342	·799242	1·251184	·7811574	22
39	·6245614	·799719	1·250438	·7809757	21
40	·6247885	·800196	1·249693	·7807940	20
41	·6250156	·800673	1·248948	·7806123	19
42	·6252427	·801151	1·248204	·7804304	18
43	·6254696	·801628	1·247460	·7802485	17
44	·6256966	·802106	1·246716	·7800665	16
45	·6259235	·802584	1·245974	·7798845	15
46	·6261503	·803063	1·245232	·7797024	14
47	·6263771	·803541	1·244490	·7795202	13
48	·6266038	·804020	1·243749	·7793380	12
49	·6268305	·804499	1·243008	·7791557	11
50	·6270571	·804979	1·242268	·7789733	10
51	·6272837	·805458	1·241529	·7787909	9
52	·6275102	·805938	1·240790	·7786084	8
53	·6277366	·806418	1·240051	·7784258	7
54	·6279631	·806898	1·239313	·7782431	6
55	·6281894	·807378	1·238576	·7780604	5
56	·6284157	·807859	1·237839	·7778777	4
57	·6286420	·808340	1·237103	·7776949	3
58	·6288682	·808821	1·236367	·7775120	2
59	·6290943	·809302	1·235631	·7773290	1
60	·6293204	·809784	1·234897	·7771460	0
′	COSINE.	COTANG.	TANG.	SINE.	′

Deg. 51.

NATURAL SINES AND TANGENTS TO A RADIUS 1.

39 Deg. | 39 Deg.

′	SINE.	TANG.	COTANG.	COSINE.	′	′	SINE.	TANG.	COTANG.	COSINE.	′
0	·6293204	·809784	1·234897	·7771460	60	31	·6363026	·824825	1·212378	·7714395	29
1	·6295464	·810265	1·234162	·7769629	59	32	·6365270	·825314	1·211660	·7712544	28
2	·6297724	·810747	1·233429	·7767797	58	33	·6367513	·825803	1·210942	·7710692	27
3	·6299983	·811230	1·232696	·7765965	57	34	·6369756	·826292	1·210225	·7708840	26
4	·6302242	·811712	1·231963	·7764132	56	35	·6371998	·826782	1·209508	·7706986	25
5	·6304500	·812195	1·231231	·7762298	55	36	·6374240	·827271	1·208792	·7705132	24
6	·6306758	·812678	1·230499	·7760464	54	37	·6376481	·827762	1·208076	·7703278	23
7	·6309015	·813161	1·229768	·7758629	53	38	·6378721	·828252	1·207361	·7701423	22
8	·6311272	·813644	1·229038	·7756794	52	39	·6380961	·828742	1·206646	·7699567	21
9	·6313528	·814128	1·228308	·7754957	51	40	·6383201	·829233	1·205932	·7697710	20
10	·6315784	·814611	1·227578	·7753121	50	41	·6385440	·829724	1·205219	·7695853	19
11	·6318039	·815095	1·226849	·7751283	49	42	·6387678	·830216	1·204505	·7693996	18
12	·6320293	·815580	1·226121	·7749445	48	43	·6389916	·830707	1·203793	·7692137	17
13	·6322547	·816064	1·225393	·7747606	47	44	·6392153	·831199	1·203081	·7690278	16
14	·6324800	·816549	1·224665	·7745767	46	45	·6394390	·831691	1·202369	·7688418	15
15	·6327053	·817034	1·223938	·7743926	45	46	·6396626	·832183	1·201658	·7686558	14
16	·6329306	·817519	1·223212	·7742086	44	47	·6398862	·832675	1·200947	·7684697	13
17	·6331557	·818004	1·222486	·7740244	43	48	·6401097	·833168	1·200237	·7682835	12
18	·6333809	·818490	1·221761	·7738402	42	49	·6403332	·833661	1·199527	·7680973	11
19	·6336059	·818976	1·221036	·7736559	41	50	·6405566	·834154	1·198818	·7679110	10
20	·6338310	·819462	1·220312	·7734716	40	51	·6407799	·834648	1·198109	·7677246	9
21	·6340559	·819948	1·219588	·7732872	39	52	·6410032	·835141	1·197401	·7675382	8
22	·6342808	·820435	1·218865	·7731027	38	53	·6412264	·835635	1·196693	·7673517	7
23	·6345057	·820922	1·218142	·7729182	37	54	·6414496	·836129	1·195986	·7671652	6
24	·6347305	·821409	1·217419	·7727336	36	55	·6416728	·836624	1·195279	·7669785	5
25	·6349553	·821896	1·216698	·7725489	35	56	·6418958	·837118	1·194573	·7667918	4
26	·6351800	·822384	1·215976	·7723642	34	57	·6421189	·837613	1·193867	·7666051	3
27	·6354046	·822871	1·215256	·7721794	33	58	·6423418	·838108	1·193162	·7664183	2
28	·6356292	·823359	1·214535	·7719945	32	59	·6425647	·838604	1·192457	·7662314	1
29	·6358537	·823847	1·213816	·7718096	31	60	·6427876	·839099	1·191753	·7660444	0
30	·6360782	·824336	1·213097	·7716246	30						
′	COSINE.	COTANG.	TANG.	SINE.	′	′	COSINE.	COTANG.	TANG.	SINE.	′

Deg. 50. | Deg. 50.

NATURAL SINES AND TANGENTS TO A RADIUS 1.

40 Deg.

′	SINE.	TANG.	COTANG.	COSINE.	′
0	·6427876	839099	1·191753	·7660444	60
1	·6430104	839595	1·191049	·7658574	59
2	·6432332	840091	1·190346	·7656704	58
3	·6434559	840587	1·189643	·7654832	57
4	·6436785	841084	1·188941	·7652960	56
5	·6439011	841581	1·188239	·7651087	55
6	·6441236	842078	1·187538	·7649214	54
7	·6443461	842575	1·186837	·7647340	53
8	·6445685	843073	1·186136	·7645465	52
9	·6447909	843570	1·185437	·7643590	51
10	·6450132	844068	1·184737	·7641714	50
11	·6452355	844567	1·184038	·7639838	49
12	·6454577	845065	1·183340	·7637960	48
13	·6456798	845564	1·182642	·7636082	47
14	·6459019	846063	1·181944	·7634204	46
15	·6461240	846562	1·181247	·7632325	45
16	·6463460	847062	1·180551	·7630445	44
17	·6465679	847561	1·179855	·7628564	43
18	·6467898	848061	1·179159	·7626683	42
19	·6470116	848561	1·178464	·7624802	41
20	6472334	849062	1·177769	·7622919	40
21	6474551	849563	1·177075	·7621036	39
22	·6476767	850064	1·176382	·7619152	38
23	·6478984	·850565	1·175688	·7617268	37
24	·6481199	·851066	1·174996	·7615383	36
25	·6483414	·851568	1·174303	·7613497	35
26	·6485628	·852070	1·173612	·7611611	34
27	·6487842	852572	1·172920	·7609724	33
28	·6490056	853075	1·172229	·7607837	32
29	6492268	853577	1·171539	·7605949	31
30	·6494480	·854080	1·170849	·7604060	30
′	COSINE.	COTANG.	TANG.	SINE.	′

Deg. 49.

40 Deg.

′	SINE.	TANG.	COTANG.	COSINE.	′
31	·6496692	·854583	1·170160	·7602170	29
32	·6498903	·855087	1·169471	·7600280	28
33	·6501114	·855591	1·168782	·7598389	27
34	·6503324	·856095	1·168094	·7596498	26
35	·6505533	·856599	1.167407	·7594606	25
36	·6507742	·857103	1·166720	·7592713	24
37	·6509951	·857608	1·166033	·7590820	23
38	·6512158	·858113	1·165347	·7588926	22
39	·6514366	·858618	1·164661	·7587031	21
40	·6516572	·859124	1·163976	·7585136	20
41	·6518778	·859629	1·163291	·7583240	19
42	·6520984	·860135	1·162607	·7581343	18
43	·6523189	·860641	1·161923	·7579446	17
44	·6525394	·861148	1·161240	·7577548	16
45	·6527598	·861655	1·160557	·7575650	15
46	·6529801	·862162	1·159874	·7573751	14
47	·6532004	·862669	1·159192	·7571851	13
48	·6534206	·863176	1·158511	·7569951	12
49	·6536408	·863684	1·157830	·7568050	11
50	·6538609	·864192	1·157149	·7566148	10
51	·6540810	·864700	1·156469	·7564246	9
52	·6543010	·865209	1·155789	·7562343	8
53	·6545209	·865718	1·155110	·7560439	7
54	·6547408	·866227	1·154431	·7558535	6
55	·6549607	·866736	1·153753	·7556630	5
56	·6551804	·867246	1·153075	·7554724	4
57	·6554002	·867755	1·152397	·7552818	3
58	·6556198	·868265	1·151721	·7550911	2
59	·6558395	·868776	1·151044	·7549004	1
60	·6560590	·869286	1·150368	7547096	0
′	COSINE.	COTANG.	TANG.	SINE.	′

Deg. 49.

NATURAL SINES AND TANGENTS TO A RADIUS 1.

41 Deg.

′	SINE.	TANG.	COTANG.	COSINE.	′
0	·6560590	·869286	1·150368	·7547096	60
1	·6562785	·869797	1·149692	·7545187	59
2	·6564980	·870308	1·149017	·7543278	58
3	·6567174	·870820	1·148342	·7541368	57
4	·6569367	·871331	1·147668	·7539457	56
5	·6571560	·871843	1·146994	·7537546	55
6	·6573752	·872355	1·146321	·7535634	54
7	·6575944	·872868	1·145648	·7533721	53
8	·6578135	·873380	1·144976	·7531808	52
9	·6580326	·873893	1·144304	·7529894	51
10	·6582516	·874406	1·143632	·7527980	50
11	·6584706	·874920	1·142961	·7526065	49
12	·6586895	·875433	1·142290	·7524149	48
13	·6589083	·875947	1·141620	·7522233	47
14	·6591271	·876462	1·140950	·7520316	46
15	·6593458	·876976	1·140281	·7518398	45
16	·6595645	·877491	1·139612	·7516480	44
17	·6597831	·878006	1·138944	·7514561	43
18	·6600017	·878521	1·138276	·7512641	42
19	·6602202	·879037	1·137608	·7510721	41
20	·6604386	·879552	1·136941	·7508800	40
21	·6606570	·880068	1·136274	·7506879	39
22	·6608754	·880585	1·135608	·7504957	38
23	·6610936	·881101	1·134942	·7503034	37
24	·6613119	·881618	1·134277	·7501111	36
25	·6615300	·882135	1·133612	·7499187	35
26	·6617482	·882653	1·132947	·7497262	34
27	·6619662	·883170	1·132283	·7495337	33
28	·6621842	·883688	1·131620	·7493411	32
29	·6624022	·884206	1·130957	·7491484	31
30	·6626200	·884725	1·130294	·7489557	30
′	COSINE.	COTANG.	TANG.	SINE.	′

Deg. 48.

41 Deg.

′	SINE.	TANG.	COTANG.	COSINE.	′
31	·6628379	·885244	1·129632	·7487629	29
32	·6630557	·885763	1·128970	·7485701	28
33	·6632734	·886282	1·128308	·7483772	27
34	·6634910	·886801	1·127647	·7481842	26
35	·6637087	·887321	1.126987	·7479912	25
36	·6639262	·887841	1·126327	·7477981	24
37	·6641437	·888361	1·125667	·7476049	23
38	·6643612	·888882	1·125008	·7474117	22
39	·6645785	·889403	1·124349	·7472184	21
40	·6647959	·889924	1·123690	·7470251	20
41	·6650131	·890445	1·123032	·7468317	19
42	·6652304	·890967	1·122375	·7466382	18
43	·6654475	·891489	1·121718	·7464446	17
44	·6656646	·892011	1·121061	·7462510	16
45	·6658817	·892534	1·120405	·7460574	15
46	·6660987	·893056	1·119749	·7458636	14
47	·6663156	·893579	1·119094	·7456699	13
48	·6665325	·894103	1·118439	·7454760	12
49	·6667493	·894626	1·117784	·7452821	11
50	·6669661	·895150	1·117130	·7450881	10
51	·6671828	·895674	1·116476	·7448941	9
52	·6673994	·896199	1·115823	·7446999	8
53	·6676160	·896723	1·115170	·7445058	7
54	·6678326	·897248	1·114518	·7443115	6
55	·6680490	·897773	1·113866	·7441173	5
56	·6682655	·898299	1·113214	·7439229	4
57	·6684818	·898825	1·112563	·7437285	3
58	·6686981	·899351	1·111912	·7435340	2
59	·6689144	·899877	1·111262	·7433394	1
60	·6691306	·900404	1·110612	·7431448	0
′	COSINE.	COTANG.	TANG.	SINE.	′

Deg. 48.

NATURAL SINES AND TANGENTS TO A RADIUS 1.

42 Deg. | 42 Deg.

′	SINE.	TANG.	COTANG.	COSINE.	′	′	SINE.	TANG.	COTANG.	COSINE	′
0	6691306	·900404	1·110612	·7431448	60	31	·6758046	·916866	1·090671	·7370808	29
1	·6693468	·900930	1·109963	·7429502	59	32	·6760190	·917402	1·090034	·7368842	28
2	·6695628.	·901458	1·109314	·7427554	58	33	·6762333	·917937	1·089398	·7366875	27
3	·6697789	·901985	1·108665	·7425606	57	34	·6764476	·918474	1·088762	·7364908	26
4	·6699948	·902513	1·108017	·7423658	56	35	·6766618	·919010	1.088126	·7362940	25
5	·6702108	·903041	1·107369	·7421708	55	36	·6768760	·919547	1·087491	·7360971	24
6	·6704266	·903569	1·106721	·7419758	54	37	·6770901	·920084	1·086857	·7359002	23
7	·6706424	·904097	1·106075	·7417808	53	38	·6773041	·920621	1·086222	·7357032	22
8	·6708582	·904626	1·105428	·7415857	52	39	·6775181	·921159	1·085588	·7355061	21
9	·6710739	·905155	1·104782	·7413905	51	40	·6777320	·921696	1·084955	·7353090	20
10	·6712895	·905685	1·104136	·7411953	50	41	·6779459	·922235	1·084322	·7351118	19
11	·6715051	·906214	1·103491	·7410000	49	42	·6781597	·922773	1·083689	·7349146	18
12	·6717206	·906744	1·102846	·7408046	48	43	·6783734	·923312	1·083057	·7347173	17
13	·6719361	·907274	1·102201	·7406092	47	44	·6785871	·923851	1·082425	·7345199	16
14	·6721515	·907805	1·101557	·7404137	46	45	·6788007	·924390	1·081793	·7343225	15
15	·6723668	·908336	1·100914	·7402181	45	46	·6790143	·924930	1·081162	·7341250	14
16	·6725821	·908867	1·100270	·7400225	44	47	·6792278	·925470	1·080532	·7339275	13
17	·6727973	·909398	1·099628	·7398268	43	48	·6794413	·926010	1·079901	·7337299	12
18	·6730125	·909930	1·098985	·7396311	42	49	·6796547	·926550	1·079271	·7335322	11
19	·6732276	·910461	1·098343	·7394353	41	50	·6798681	·927091	1·078642	·7333345	10
20	·6734427	·910994	1·097702	·7392394	40	51	·6800813	·927632	1·078013	·7331367	9
21	·6736577	·911526	1·097060	·7390435	39	52	·6802946	·928173	1·077384	·7329388	8
22	·6738727	·912059	1·096420	·7388475	38	53	·6805078	·928715	1·076756	·7327409	7
23	·6740876	·912592	1·095779	·7386515	37	54	·6807209	·929257	1·076128	·7325429	6
24	·6743024	·913125	1·095139	·7384553	36	55	·6809339	·929799	1·075500	·7323449	5
25	·6745172	·913659	1·094500	·7382592	35	56	·6811469	·930342	1·074873	·7321467	4
26	·6747319	·914192	1·093861	·7380629	34	57	·6813599	·930884	1·074246	·7319486	3
27	·6749466	·914727	1·093222	·7378666	33	58	·6815728	·931428	1·073620	·7317503	2
28	·6751612	·915261	1·092584	·7376703	32	59	·6817856	·931971	1·072994	·7315521	1
29	·6753757	·915796	1·091946	·7374738	31	60	·6819984	·932515	1·072368	·7313537	0
30	·6755902	·916331	1·091308	·7372773	30						
′	COSINE.	COTANG.	TANG.	SINE.	′	′	COSINE.	COTANG.	TANG.	SINE.	′

Deg. 47. | Deg. 47.

NATURAL SINES AND TANGENTS TO A RADIUS 1.

43 Deg. | 43 Deg.

′	SINE.	TANG.	COTANG.	COSINE.	′	′	SINE.	TANG.	COTANG.	COSINE.	′
0	·6819984	·932515	1·072368	·7313537	60	31	·6885655	·949517	1·053166	·7251741	29
1	·6822111	·933059	1·071743	·7311553	59	32	·6887765	·950070	1·052553	·7249738	28
2	·6824237	·933603	1·071118	·7309568	58	33	·6889873	·950624	1·051940	·7247734	27
3	·6826363	·934147	1·070494	·7307583	57	34	·6891981	·951178	1·051327	·7245729	26
4	·6828489	·934692	1·069870	·7305597	56	35	·6894089	·951732	1·050715	·7243724	25
5	·6830613	·935238	1·069246	·7303610	55	36	·6896195	·952287	1·050103	·7241719	24
6	·6832738	·935783	1·068623	·7301623	54	37	·6898302	·952842	1·049492	·7239712	23
7	·6834861	·936329	1·068000	·7299635	53	38	·6900407	·953397	1·048880	·7237705	22
8	·6836984	·936875	1·067377	·7297646	52	39	·6902512	·953952	1·048270	·7235698	21
9	·6839107	·937421	1·066755	·7295657	51	40	·6904617	·954508	1·047659	·7233690	20
10	·6841229	·937968	1·066134	·7293668	50	41	·6906721	·955064	1·047049	·7231681	19
11	·6843350	·938515	1·065512	·7291677	49	42	·6908824	·955620	1·046440	·7229671	18
12	·6845471	·939062	1·064891	·7289686	48	43	·6910927	·956177	1·045831	·7227661	17
13	·6847591	·939610	1·064271	·7287695	47	44	·6913029	·956734	1·045222	·7225651	16
14	·6849711	·940157	1·063651	·7285703	46	45	·6915131	·957291	1·044613	·7223640	15
15	·6851830	·940706	1·063031	·7283710	45	46	·6917232	·957849	1·044005	·7221628	14
16	·6853948	·941254	1·062411	·7281716	44	47	·6919332	·958407	1·043397	·7219615	13
17	·6856066	·941803	1·061792	·7279722	43	48	·6921432	·958965	1·042790	·7217602	12
18	·6858184	·942352	1·061174	·7277728	42	49	·6923531	·959524	1·042183	·7215589	11
19	·6860300	·942901	1·060556	·7275732	41	50	·6925630	·960082	1·041576	·7213574	10
20	·6862416	·943451	1·059938	·7273736	40	51	·6927728	·960642	1·040970	·7211559	9
21	·6864532	·944001	1·059320	·7271740	39	52	·6929825	·961201	1·040364	·7209544	8
22	·6866647	·944551	1·058703	·7269743	38	53	·6931922	·961761	1·039758	·7207528	7
23	·6868761	·945102	1·058086	·7267745	37	54	·6934018	·962321	1·039153	·7205511	6
24	·6870875	·945653	1·057470	·7265747	36	55	·6936114	·962881	1·038548	·7203494	5
25	·6872988	·946204	1·056854	·7263748	35	56	·6938209	·963442	1·037944	·7201476	4
26	·6875101	·946755	1·056238	·7261748	34	57	·6940304	·964003	1·037340	·7199457	3
27	·6877213	·947307	1·055623	·7259748	33	58	·6942398	·964565	1·036736	·7197438	2
28	·6879325	·947859	1·055008	·7257747	32	59	·6944491	·965126	1·036133	·7195418	1
29	·6881435	·948411	1·054394	·7255746	31	60	·6946584	·965688	1·035530	·7193398	0
30	·6883546	·948964	1·053780	·7253744	30						
	COSINE.	COTANG.	TANG.	SINE.	′	′	COSINE.	COTANG.	TANG.	SINE.	′

Deg 46. | Deg. 46.

NATURAL SINES AND TANGENTS TO A RADIUS 1.

44 Deg. | 44 Deg.

′	SINE.	TANG.	COTANG.	COSINE.	′	′	SINE.	TANG.	COTANG.	COSINE.	′
0	·6946584	·965688	1·035530	·7193398	60	31	·7011167	·983269	1·017015	·7130465	29
1	·6948676	·966251	1·034927	·7191377	59	32	·7013241	·983841	1·016423	·7128426	28
2	·6950767	·966813	1·034325	·7189355	58	33	·7015314	·984414	1·015832	·7126385	27
3	·6952858	·967376	1·033723	·7187333	57	34	·7017387	·984987	1·015241	·7124344	26
4	·6954949	·967939	1·033122	·7185310	56	35	·7019459	·985560	1·014651	·7122303	25
5	·6957039	·968503	1·032520	·7183287	55	36	·7021531	·986133	1·014061	·7120260	24
6	·6959128	·969067	1·031919	·7181263	54	37	·7023601	·986707	1·013471	·7118218	23
7	·6961217	·969631	1·031319	·7179238	53	38	·7025672	·987282	1·012881	·7116174	22
8	·6963305	·970196	1·030719	·7177213	52	39	·7027741	·987856	1·012292	·7114130	21
9	·6965392	·970761	1·030119	·7175187	51	40	·7029811	·988431	1·011703	·7112086	20
10	·6967479	·971326	1·029520	·7173161	50	41	·7031879	·989006	1·011115	·7110041	19
11	·6969565	·971891	1·028921	·7171134	49	42	·7033947	·989582	1·010527	·7107995	18
12	·6971651	·972457	1·028322	·7169106	48	43	·7036014	·990158	1·009939	·7105948	17
13	·6973736	·973023	1·027724	·7167078	47	44	·7038081	·990734	1·009352	·7103901	16
14	·6975821	·973590	1·027126	·7165049	46	45	·7040147	·991311	1·008764	·7101854	15
15	·6977905	·974156	1·026528	·7163019	45	46	·7042213	·991888	1·008178	·7099806	14
16	·6979988	·974724	1·025931	·7160989	44	47	·7044278	·992465	1·007591	·7097757	13
17	·6982071	·975291	1·025334	·7158959	43	48	·7046342	·993042	1·007005	·7095707	12
18	·6984153	·975859	1·024738	·7156927	42	49	·7048406	·993620	1·006420	·7093657	11
19	·6986234.	·976427	1·024141	·7154895	41	50	·7050469	·994199	1·005834	·7091607	10
20	·6988315	·976995	1·023546	·7152863	40	51	·7052532	·994777	1·005249	·7089556	9
21	·6990396	·977564	1·022950	·7150830	39	52	·7054594	·995356	1·004665	·7087504	8
22	·6992476	·978133	1·022355	·7148796	38	53	·7056655	·995935	1·004080	·7085451	7
23	·6994555	·978702	1·021760	·7146762	37	54	·7058716	·996515	1·003496	·7083398	6
24	·6996633	·979272	1·021166	·7144727	36	55	·7060776	·997095	1·002913	·7081345	5
25	·6998711	·979842	1·020572	·7142691	35	56	·7062835	·997675	1·002329	·7079291	4
26	·7000789	·980412	1·019978	·7140655	34	57	·7064894	·998256	1·001746	·7077236	3
27	·7002866	·980983	1·019385	·7138618	33	58	·7066953	·998837	1·001164	·7075180	2
28	·7004942	·981554	1·018792	·7136581	32	59	·7069011	·999418	1·000581	·7073124	1
29	·7007018	·982125	1·018199	·7134543	31	60	·7071068	1·00000	1·000000	·7071068	0
30	·7009093	·982697	1·017607	·7132504	30						
′	COSINE.	COTANG.	TANG.	SINE.	′	′	COSINE.	COTANG.	TANG.	SINE.	′

Deg. 45. | Deg. 45.

www.ingramcontent.com/pod-product-compliance
Lightning Source LLC
LaVergne TN
LVHW021413110826
845150LV00007B/1903

* 9 7 8 1 4 2 5 5 1 0 7 0 1 *